The Power Of
VEDIC
MATHS

The Power Of
VEDIC
MATHS

For Admission Tests, Professional & Competitive Examinations

Atul Gupta

JAICO PUBLISHING HOUSE

Mumbai • Delhi • Bangalore • Kolkata
Hyderabad • Chennai • Ahmedabad • Bhopal

Published by Jaico Publishing House
121 Mahatma Gandhi Road
Mumbai - 400 023
jaicopub@vsnl.com
www.jaicobooks.com

THE POWER OF VEDIC MATHS
ISBN 81-7992-357-6

First Jaico Impression: 2004
Fourth Jaico Impression: 2005

Printed by
Snehesh Printers
320-A, Shah & Nahar Ind. Est. A-1
Lower Parel, Mumbai - 400 013.

Contents

Preface

Mathematics is considered to be a dry and boring subject by a large number of people. Children dislike and fear mathematics for a variety of reasons. This book is written with the sole purpose of helping school and college students, teachers, parents, common people and people from non-mathematical areas of study, to discover the joys of solving mathematical problems, by a wonderful set of techniques called 'Vedic Maths'.

These techniques are derived from 16 sutras (verses) in the Vedas, which are thousands of years old and among the earliest literature of ancient Hindus in India. They are an endless source of knowledge and wisdom, providing practical knowledge in all spheres of life. Jagadguru Swami Sri Bharati Krshna Tirthaji (1884-1960), was a brilliant scholar, who discovered the 16 sutras in the Vedas and spent 8 years in their intense study. He has left an invaluable treasure for all generations to come, consisting of a set of unique and magnificent methods for solving mathematical problems in areas like arithmetic, algebra, calculus, trigonometry and co-ordinate geometry. These techniques are very easy to learn and encapsulate the immense and brilliant mathematical knowledge of ancient Indians, who had made fundamental contributions to mathematics in the form of the decimal numerals, zero and infinity.

I have trained thousands of children of all age groups with these techniques and I find that even young children enjoy learning and using them. The techniques reduce drastically, the number of steps required to solve problems and in many cases, after a little practice, many of the problems can be solved orally. It gives tremendous self-confidence to the students which leads them to enjoy mathematics instead of fearing and disliking it.

I have written this book in the form of a cookbook, where a reader can grasp a technique quickly, instead of reading through a large mass of theory before understanding it. I have considered techniques for major arithmetical operations like multiplication, division, computation of squares and square roots and complex fractions, besides a whole lot of other techniques. Each technique has been explained in detail with the help of solved examples, using a step-by-step approach, and I am sure that the reader will be able to understand and master the contents easily. Every chapter has a large number of problems for practice and the book contains over 1000 problems for practice. The answers are given alongside so that the reader can either solve the problems orally or use paper and pen and compare with the given answer. The chapters should be read sequentially to absorb the material and then can be used for reference in any desired order.

I have also included a special chapter, in which I have shown the application of the techniques to solve problems, collected from several competitive exams. This is a unique feature of the book and should add to the popularity of the techniques.

We have tried to make all the examples and answers error-free but if any mistake is discovered, we will be obliged if we are informed about the same.

Constructive criticism and comments can be sent to the author at gupta@bom3.vsnl.net.in.

ATUL GUPTA

CHAPTER 1

Two Simple Techniques

We will begin our journey into the fascinating world of vedic maths with two simple techniques which will lay the foundation for some of the techniques in the following chapters.

I. Subtraction from 100/1000/10000

We will start with a very simple technique wherein we will see the use of the sutra 'All from 9 and last from 10'. This is used to subtract a given number from 100, 1000, 10000 etc. It removes the mental strain which is existent in the method taught in schools.

This method is also used later on in the Nikhilam method of multiplication.

Consider the subtraction of 7672 from 10000.

a) Normal Method

The normal method is

```
    1 0 0 0 0
  -   7 6 7 2
  _____
    0 2 3 2 8
```

We carry '1' from the left side and continue doing so till we reach the right-most digit, leaving behind 9 in each column and 10 in the last column.

Then, we subtract the right-most digit '2' from 10 and write down '8'. Next, we subtract the digit '7' from '9' and write down '2'.

We repeat this process for all the remaining digits to the left.

Through this operation, the final result is always obtained from right to left.

Mentally, there is a carry operation for every digit, which is time consuming and slows down the overall process.

b) Vedic Method

The vedic method uses the sutra "All from 9 and last from 10" and gives a very simple and powerful technique to achieve the same result.

The result can be obtained from both left to right as well as right to left with equal ease. It states that the result can be obtained by subtraction of each digit from '9' and the last digit from '10'.

Hence, in the given example,

$$
\begin{array}{r}
1\,0\,0\,0\,0 \\
-\quad 7\,6\,7\,2 \\
\hline
\end{array}
$$

We can get the result from left to right or vice versa from right to left as

(9 – 7)	(9 – 6)	(9 – 7)	(10 – 2)
2	3	2	8

i.e. all digits except the last one are subtracted from 9, the last digit is subtracted from 10 and the result (2328) is written down directly. The mental burden of a carry for each column vanishes and the answer can be obtained easily, in a jiffy.

The same technique can be applied for decimal subtraction also, e.g.   2.000 − 0.3436.

The core operation here is subtraction of 3436 from 10000 where 1 is a carry from left.

1. *Examples of subtractions from bases of 100, 1000 etc.*

Sr No	Example	Answer
1.	100 − 37	63
2.	1000 − 293	707
3.	100 − 81	19
4.	1000 − 889	111
5.	10000 − 3725	6275
6.	10000 − 8889	1111
7.	100000 − 5328	94672
8.	100 − 63	37
9.	1000 − 63	937
10.	10000 − 362	9638
11.	1.000 − 0.764	0.236
12.	2.0 − 0.672	1.328
13.	3.0 − 1.282	1.718
14.	10.00 − 0.89	9.11
15.	80.000 − 0.7312	79.2688
16.	98.000 − 3.255	94.745
17.	2.00 − 1.3628	0.6372

18.	55.00 − 14.3827	40.6173
19.	277000 − 277	276723
20.	500000 − 433	499567

This simple subtraction method will be used further in multiplication (Ch.4) and mishrank (Ch.12).

II. Multiplication with a series of 9s

a) Multiplication of a number by same number of 9s

Let us use this technique to see how to carry out multiplication of a n-digit number by a number with 'n' number of 9s eg. any 3-digit number by 999 or a 4-digit number by 9999.

Let us see an example viz   533 × 999

The solution can be obtained as 533 × (1000 − 1)
$$= 533000 - 533$$
$$= 532467.$$

The answer consists of two 3-digit numbers viz '532' and '467'

The first part '532' is 1 less than the given number, i.e. 533 − 1.

The second part '467' is equal to (1000 − 533) or (999 − 532). It is simply the '9s' complement of the first 3 digits of the result; i.e. 9s complement of 532 where the required digits can be obtained as:

$$4 = 9 - 5$$
$$6 = 9 - 3$$
$$7 = 9 - 2$$

Similarly, 3456 × 9999 will consist of two parts

3455 / 6544

i.e. 34556544 (where, 6544 is '9's complement of 3455).

This simple technique can be used to get the result orally whenever any number is multiplied by a number consisting of an equal number of 9s.

The same technique can also be used for multiplication of decimal numbers.

Consider the example 4.36 × 99.9

Firstly, multiply 436 by 999 and write down the result as 435564.

Now, since the numbers have two and one decimal places respectively, the final result will have three decimal places and we put the decimal point three points from the right. So, the final answer will be 435.564.

b) Multiplication of a number by greater number of 9s

Let us now see how to carry out multiplication of a n-digit number by a number with greater number of '9's
Eg. 235 by 9999 or 235 by 99999

i) Consider 235 × 9999

Since the number of digits in 235 is three while there are four digits in 9999, we pad 235 with one zero to get 0235 and carry out the multiplication as before.

0235 × 9999 = 0234/9765 and the final result is 2349765.

ii) Consider 235 × 99999

Padding 235 with two zeroes, we get
    00235 × 99999 = 00234/99765
and the final result is 23499765.

c) Multiplication of a number by lesser number of 9s

Let us now see how to carry out multiplication of a n-digit number by a number with lesser number of 9s.

i) Consider 235 × 99

Since there are two digits in 99, we place a colon two digits from the right in the given number to get
    2 : 35

We enclose this number within a colon on the right to get
    2 : 35 :

We now increase the number to the left of the colon by 1 and derive a new number as
    3 : 35

We align the colon of this new number with the enclosing colon of the original number to get
    2 : 35 :
        3 : 35

We now subtract 3 from 235 and 35 from 100, to get the final answer as

$$\begin{array}{r} 2 : 35 : \\ 3 \; : 35 \\ \hline 2 \quad 32 \quad 65 \end{array}$$

ii) *Consider 12456 × 99*

Since there are two digits in 99, we place a colon two digits from the right in the given number to get

    124 : 56

We enclose this number within a colon on the right to get

    124 : 56 :

We now increase the number to the left of the colon by 1 and derive a new number as

    125 : 56

We align the colon of this new number with the enclosing colon of the original number to get

    124   : 56 :
      1   25 : 56

We now subtract to get the final answer as

    124   : 56 :
      1   25 : 56
    —————————————
    123   31   44

2. *Examples for multiplication by series of 9s*

Sr No	Example	Answer
1.	253 × 999	252747
2.	680 × 999	679320
3.	7238 × 9999	72372762
4.	8000 × 9999	79992000
5.	72.81 × 9.999	728.02719
6.	35152 × 99999	3515164848
7.	1.785 × 999.9	1784.8215

8.	0.0055 × 0.99	0.005445
9.	8.88 × 99.9	887.112
10.	232 × 9999	2319768
11.	365.7 × 99999	36569634.30
12.	8.91 × 999.9	8909.109
13.	7856 × 999	7848144
14.	123.5 × 999990	123498765
15.	999 × 99900	99800100
16.	2.69 × 999	2687.31
17.	76.4 × 99.9	7632.36
18.	781 × 99	77319
19.	143.5 × 999.9	143485.65
20.	0.16 × 0.99	0.1584

In examples, where decimal numbers are involved, we multiply the two numbers assuming that there are no decimals and then place the decimal in the right place, as explained above.

CHAPTER 2

Operations with 9

The number 9 has a special significance in Vedic Maths. We will start by looking at a simple but powerful technique which is used to get the remainder on dividing any number by 9. This remainder is given several names like 'Navasesh', 'Beejank' or digital root. It has powerful properties which are used to check the correctness of arithmetic operations like addition, subtraction, multiplication, squaring, cubing etc. It can also be used to check the divisibility of any given number by 9.

We will begin by seeing how to compute the navasesh of any given number and then proceed to see its applications.

I. Computation of remainder (Navasesh) on division by 9

The navasesh of a number is defined as the remainder obtained on dividing that number by 9.

Example :   The navasesh of 20 is 2
The navasesh of 181 is 1
The navasesh of 8132 is 5
The navasesh of 2357 is 8

Let us see how to use techniques provided by Vedic maths to compute the navasesh for any number. We will denote it by 'N'.

a) Basic Method

To compute the navasesh (N) add all the digits in the number.

- If we get a single digit as the answer, then that is the navasesh (if the sum is 0 or 9, then the navasesh is 0).

- If there is more than one digit in the sum, repeat the process on the sum obtained. Keep repeating till a single digit is obtained.

Examples :

Navasesh of 20 can be computed by adding the digits 2 and 0 to get the result as 2.

Navasesh of 8132 is similarly obtained as follows :
    Pass 1 : 8 + 1 + 3 + 2 gives 14,
which has more than one digit.
    Pass 2 : 1 + 4 gives 5
which is the required navasesh.

b) First enhancement

If there is one or more '9's in the original number, these need not be added and can be ignored while computing the sum of the digits.

Example : If the number is 23592, then 'N' can be computed as follows.
            Pass 1 : 2 + 3 + 5 + 2 = 12 (ignore 9)
            Pass 2 : 1 + 2 = 3.
So the navasesh is 3.

c) Second enhancement

While summing up the digits, ignore groups which add up to 9.

Example : If the number is 23579, then 'N' can be computed as follows:

Ignore '9' as explained above

Ignore 2 + 7 (i.e. 1st and 4th digits which add up to '9') and add only '3' and '5' to get the 'N' as 8..

d) Final Compact Method

As before, ignore all '9's and groups which add up to 9.

While summing up the remaining digits, if the sum adds up to more than two digits at any intermediate point, reduce it to one digit there itself, by adding up the two digits. This will eliminate the need for the second pass shown in the examples above.

Example : If the number is 44629, then 'N' can be computed as follows:

Ignore '9' in the last position.

Add the remaining digits from the left where the first three digits, viz 4 + 4 + 6 add upto 14.

Since, this has two digits, we add these digits here itself to get a single digit,   i.e. 1 + 4 = 5

Now add '5' to the remaining digit '2' to get the 'N' as 7.

All methods give the same answer, but the result can be obtained faster and with lesser computation when we use the improved methods.

II) Verification of the product of two numbers

The correctness of any arithmetic operation can be verified by carrying out the same operation on the navasesh of the numbers in the operation. Once we have learnt how to compute the navasesh of a number, we can use it to check whether the result of operations like multiplication, addition and subtraction on two (or more) numbers is correct or not.

Let us see an example.

Let's take the product of 38 × 53 which is 2014.

How do we verify the correctness of the answer '2014'?

Let's take the 'N' of each of the multiplicands and of the product and see the relation between the same.

Now,   N(38) = 2   ..   Navasesh of the first number
        N(53) = 8   ..   Navasesh of the second number
        N(2014) = 7   ..   Navasesh of the product

Consider the product of the 'N' of the 2 multiplicands i.e. N(38) × N(53) = 2 × 8 = 16 whose navasesh in turn is 7.

The 'N' of the product is also '7' !!

Thus the **product** of the navasesh of the two multiplicands is equal to the navasesh of the product itself. This simple test can be used to verify whether the answer is correct.

Another Example

22.4 × 1.81 = 40.544

On ignoring the decimal points, the navasesh of the two numbers are

N(224) = 8 and
N(181) = 1 and the product of the navasesh is 8.

The navasesh of the product is also N(40544) = 8 and hence the answer is verified. The position of the decimal point can be verified easily.

III) Verification of the sum of two numbers

Similarly, let's see the case of the sum of the two numbers 38 + 53 which is 91.

We see that N(38) + N(53) = 2 + 8 = 10 which in turn has a navasesh of 1.

The 'N' of the sum (91) is also 1 !!

Thus the **sum** of the navasesh of the two numbers is equal to the navasesh of the sum itself.

IV) Verification of the difference of two numbers

Let's see the case of the difference of the two numbers.

$$53 - 38 = 15$$
$$N(53) - N(38) = 8 - 2 = 6$$
$$N(15) = 6$$

Thus the **difference** of the navasesh of the two numbers is equal to the navasesh of the difference itself.

Sometimes, the difference of the navasesh may be negative. Then, it can be converted to a positive value by adding 9 to it.

Thus N(−5) = 4, N(−6) = 3.

Let us take another example, viz

$$65 - 23 = 42$$
$$N(65) = 2,$$
$$N(23) = 5$$
$$N(65) - N(23) = -3,$$

As explained above, N(−3) = 6

The navasesh of the result should be 6.

We see that the navasesh of 42 is indeed 6.

Hence, we can use this method to verify the difference of two numbers.

V) Verification of the square or cube of two numbers

Let's take the case of squaring a number, say, 34.
$$34^2 = 1156$$

Now take the navasesh on both sides, i.e. navasesh of 34 is 7 and navasesh of 1156 is 4.

Let us square the navasesh of the left side i.e. $7^2 = 49$ whose navasesh is also 4.

Thus the square of the Navasesh of the given number is equal to the navasesh of the square. The reader can try verifying for the cube of any number by himself.

VI) Limitation in the verification process

This verification method has a limitation which has to be kept in mind.

Consider again the example of
$$38 \times 53 = 2014$$

We know that N(38) × N(53) should equal N(2014)

On the left side, we have, $2 \times 8 = 16$ which in turn gives a navasesh of 7.

Now, if the product on the RHS was assigned as 2015, its 'N' would be 8 from which we would correctly conclude that the result is wrong.

But if the answer was written as 2041 instead of 2014, the 'N' would still be 7, leading us wrongly to the conclusion that the product is correct.

So, we have to remember that this verification process using navasesh, would point out wrongly computed results with 100%

accuracy but there can be more than one result which would pass the 'correctness' test. We have to be careful about this. Inspite of this limitation, it remains a very powerful method for verification of results obtained by any of the main arithmetic operations.

VII) Computation of the quotient on division by 9

Let us now see how to get the quotient on dividing a number by 9. In vedic maths, the division is converted to a simple addition operation.

a) Method 1

The first method consists of a forward pass followed by a backward pass.

Example : Divide 8132 by 9

```
9 | 8    1    3   :   2
  |      8    9   :  12
  |_____
  | 8    9   12   :  14
```

Steps :

• Place a colon before the last digit, which shows the position separating the quotient and the remainder.

• Start from the left-most digit and bring it down as it is.

• Carry it (8) to the next column and add it to the digit in that column i.e. 1 + 8 and bring down 9.

• Carry 9 to the next column, add it to 3 and bring down 12.

• Repeat by carrying 12 to the next (last) column, add to 2 and bring down 14 as shown.

- After reaching the end, we work backwards from the right to the left by computing the remainder in the last column and by carrying the surplus digits to the left. We have to retain only one digit at each location.

- Examine the last column, which should hold the remainder. If the number is equal to or more than 9, we subtract the maximum multiples of 9 from it, which gives the quotient digit to be carried to the left, leaving the remainder behind.

- Since the number in the last column (14) contains one multiple of 9, we carry 1 to the left hand digit (12) and retain 5 (14 – 9) as final remainder.

- At the end of the forward pass, the digits looked as follows:

  8   9   12   :   14

  which now become

  8   9   13   :   5

- Now retain 3 and carry 1 to the left to get 10.

  8   10   3   :   5

- Next, retain 0 and carry 1 to the left to get 9.

  9   0   3   :   5

So the final computation would appear as follows:

8	1	3	:	2
	8	9	:	12
9	0	3	:	5

The quotient is 903 and the remainder is 5. Notice that the steps are very simple and very little mental effort is required to obtain the answer.

b) Method 2

Here, we will compute the result in the forward pass itself. During column-wise computation, if the digit in any column becomes 9 or more, we can transfer the multiples of 9 to the left and retain the remainder in the column.

In the above example, at the second column, when we get 9, we retain zero (9 – 9) in the column and carry one to the left. We then take 0 to the next column.

a)
8	1	3	:	2
	8		:	
8	9		:	

Here, we will subtract 9 from 9 to retain 0 and carry 1 to the left. On adding 1 to 8, we will get 9 and the conputaion will look as shown below.

b)
8	1	3	:	2
	8		:	
9	0		:	

c)
8	1	3	:	2
	8	0	:	3
9	0	3	:	5

This eliminates the need for the backward pass and increases the overall speed of computation.

Another Example

Consider the division of 8653 by 9

a)

8	6	5	:	3
	8		:	
8	14		:	

b)

8	6	5	:	3
	8		:	
9	5		:	

c)

8	6	5	:	3
	8	5	:	
9	5	10	:	

d)

8	6	5	:	3
	8	5	:	
9	6	1	:	

e)

8	6	5	:	3
	8	5	:	1
9	6	1	:	4

So, the quotient is 961 and the remainder is 4.

3. Examples for Navasesh

Sr No	Example	Quotient	Remainder
1.	73 / 9	8	1
2.	169 / 9	18	7
3.	176 / 9	19	5
4.	460 / 9	51	1
5.	908 / 9	100	8
6.	5711 / 9	634	5
7.	2357 / 9	261	8
8.	7875 / 9	875	0
9.	2341 / 9	260	1
10.	8276 / 9	919	5
11.	9279 / 9	1031	0
12.	16825 / 9	1869	4
13.	24321 / 9	2702	3
14.	53416 / 9	5935	1
15.	65222 / 9	7246	8
16.	232135 / 9	25792	7
17.	298179 / 9	33131	0
18.	318623 / 9	35402	5
19.	113224 / 9	12580	4
20.	599883 / 9	66653	6

CHAPTER 3

Operations with 11

In this chapter, we will consider two distinct operations with the number 11. Firstly, we will see how to multiply any number of any length by 11 and write the result, orally in a single line. Secondly, we will see how to test the divisibility of a given number by 11, i.e. whether a given number is fully divisible by 11 or not. These are useful techniques and if while solving a problem, one of the intermediate steps leads to multiplication by 11 or its multiple, this technique can be used to solve the problem quickly.

Once the technique for multiplication by 11 is mastered, multiplication of a number by 22, 33, 44 etc, all multiples of 11, can be carried out quickly by splitting the multiplicand as 11 × 2, 11 × 3 or 11 × 4. The given number can be multiplied by 11 and then the product can be multiplied further by the remaining digit, i.e. 2 or 3, as the case may be.

Multiplication

I.   Let us consider the multiplication of 153 by 11.

The traditional method is carried out as follows:

```
        1  5  3    or           1  5  3
  ×        1  1          ×          1  1
  ─────────────              ─────────────
     1  5  3  0                 1  5  3
  +     1  5  3              + 1  5  3  0
  ─────────────              ─────────────
     1  6  8  3                 1  6  8  3
```

The faster, one line method is as follows:

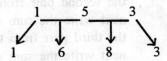

We obtain each digit of the result by adding pairs of numbers as explained below :

Steps

- Start from the right and write down the digit **3** directly

- Form pairs of consecutive digits, starting from the right and write down the sum, one at a time.

- Start from the right, form the first pair consisting of 3 and 5, add them and write down the result, i.e. write down **8**

- Move to the left, form the next pair by dropping the right-most digit 3 and including the next digit i.e. 1.

- Add 5 and 1 and write down the result **6**.

- Now, if we move left and drop the digit 5, we are left only with one digit, i.e. the left-most digit.

- Write down the left digit **1** directly.

  Thus the final answer is 1683. Isn't that real fast?

II. Let us now multiply 25345 by 11

The result is obtained step by step as shown below:

Steps

_	_	_	_	_	5	.. write down the right hand digit
_	_	_	_	9	5	.. then write the sum of '5' and '4', the first pair from the right
_	_	_	7	9	5	.. next write the sum of '4' and '3', the second pair from the right
_	_	8	7	9	5	.. next write the sum of '3' and '5', the third pair from the right
_	7	8	7	9	5	.. next write the sum of '5' and '2', the fourth pair from the right
2	7	8	7	9	5	.. once all pairs have been used up, write the left-most digit '2'

This gives the result of 25345 x 11 = 278795

III. Let's try another example : 157 x 11

We will use the same technique as follows:

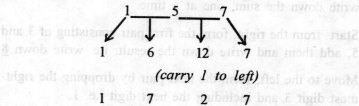

(carry 1 to left)

Here, the addition of 7 and 5 gives 12, wherein we retain 2 and carry 1 to the left. The next pair consists of 5 and 1 which on addition gives 6. Add the carry digit 1 to it to get 7.

So the final answer is 1727, which can be verified by the traditional method.

If the multiplication is between decimal numbers, we carry out the multiplication as if there are no decimals and then simply place the decimal point at the right place.

4. *Examples for Multiplication by 11*

Sr No	Exercise	Answer
1.	123 × 11	1353
2.	161 × 11	1771
3.	572 × 11	6292
4.	2135 × 11	23485
5.	7621 × 1.1	8383.10
6.	8410 × 11	92510
7.	23491 × 11	258401
8.	25200 × 11	277200
9.	7.35 × 1.1	8.085
10.	24.35 × 1.1	26.785
11.	23.48 × 0.011	0.25828
12.	53.256 × 0.11	5.85816
13.	82.4 × 0.11	9.064
14.	18.91 × 0.011	0.20801
15.	56.3 × 1.1	61.93
16.	1342 × 220	295240
17.	3215 × 330	1060950
18.	1580 × 110	173800
19.	0.00225 × 110	0.24750
20.	1342 × 22	29524
21.	4322 × 0.22	950.84
22.	141 × 0.33	46.53
23.	245 × 0.44	107.80
24.	24.4 × 55	1342
25.	546.8 × 0.55	300.74

Divisibility Test of numbers by 11

How do we test whether a number is divisible by 11?

Example : Is 23485 divisible by 11?

Steps

• Start from the right (unit's place) and add all the alternate numbers, i.e. all the digits in the odd positions viz 5, 4 and 2 which gives 11.

• Now, start from the digit in the tens place and add all the alternate numbers, i.e. all the digits in the even positions viz 8 and 3 which gives 11.

If the difference of the two sums thus obtained is either zero or divisible by 11, then the entire number is divisible by 11.
In the example above, the difference is zero and hence the number is divisible.

Let's take another example and check the divisibility of 17953.

• The sum of the digits in the odd positions is 3 + 9 + 1 which is equal to 13.

• The sum of the digits in the even positions is 5 + 7 which is equal to 12.

Since the difference of these two sums is 1 which is neither 0 nor divisible by 11, the given number is not divisible by 11.

If the given number is 17952 instead of 17953, the two sums are 12 (2 + 9 + 1) and 12 (5 + 7). The difference of the two sums is zero and hence the given number is divisible by 11.

Exercises

1. Is 8084 divisible by 11?   *Ans (No)*
2. If X381 is divisible by 11, what is the smallest value of X
                                                Ans (7)

Multiplication with 111

We can use the same technique to multiply number by 111. As before as with 11, we proceed with the computation from the right starting with 1 digit and progressively increasing the digits till we reach 3 digits. An example will make it clear.

e.g. 4132 x 111

The result is obtained step by step as shown below:

Steps

_	_	_	_	_	2	..	write down the right hand digit
_	_	_	_	5	2	..	then write the sum of '2' and '3', the first pair from the right
_	_	_	6	5	2	..	next write the sum of '2', '3' and 1, the first three digit set from the right
_	_	8	6	5	2	..	drop 2 and write the sum of '3', '1', and '4', the second three digit set from the right
_	5	8	6	5	2	..	drop three and write the sum of '1' and '4'
4	5	8	6	5	2	..	lastly, write the left-most digit '4'

This gives the result of 4132 x 111 = 458652

The technique can be extended for numbers like 1111, 11111 etc.

CHAPTER 4

Multiplication (Nikhilam)

Multiplication of numbers plays a very important role in most computations. Vedic maths offers two main approaches to multiplying 2 numbers. These are the

- Nikhilam method and
- Urdhva Tiryak method

In the Nikhilam method, we subtract a common number, called the base from both the numbers. The multiplication is then between the differences so obtained.

In the next chapter, we will take up the 'Urdhya Tiryak' method, which uses cross-multiplication.

Let us start the study of the Nikhilam method now by considering multiplication of numbers which are close to a base of 100, 1000, 10000 etc. Later on, we will see the multiplication of numbers close to bases like 50, 500, 5000 etc. and finally any convenient base.

For a base 100, the numbers which are being multiplied may be

- less than the base   (e.g. 96, 98)
- more than the base (e.g. 103, 105)
- mixed (e.g. 96 & 103)

Case 1 : Integers less than the base (100)

Consider the multiplication of 93 × 98. Let's look first at the traditional method and then at the vedic method.

```
            9   3
    ×       9   8
    _____
        8   3   7   0
    +       7   4   4
    _____
        9   1   1   4
    _____
```

Vedic Method

Write the two numbers one below the other and write the difference of the base (100) from each of the numbers, on the right side as shown :

```
    93 ⟍     ⟋  – 7  .....    (93 – 100)
    98 ⟋     ⟍  – 2  .....    (98 – 100)
    ------------------------------
    91         / 1  4
```

Take the algebraic sum of digits across any crosswise pair. eg. 93 + (–2) or 98 + (–7).

Both give the same result viz 91, which is the first part of the answer.

This is interpreted as 91 hundreds or 9100, as the base is 100. Now, just multiply the two differences (–2) × (–7) to get 14, which is the second part of the answer.

The final answer is then 9114, which can also be interpreted as 9100 + 14.

Important Points

Since the base is 100, we should have two digits (equal to number of zeroes in the base) in the second part of the answer.

•   If the number of digits is one, the number is padded with one zero on the left.

eg. 2×3 will be written as 06.

eg.

$$
\begin{array}{ll}
98 & -2 \\
97 & -3 \\
\hline
95 & /\ 06 \quad (pad\ 6\ with\ one\ zero\ to\ get\ '06') \\
\end{array}
$$

•   If the number of digits are more than two, then the left-most digit (in the hundreds position) is carried to the left.

eg.

$$
\begin{array}{ll}
88 & -12 \\
84 & -16 \\
\hline
72 & /\ 192 \\
\end{array}
$$

= 73 / 92   (carry 1 to the left and add to 72 to get 73)

•   The method is equally effective even if one number is close to the base and the other is not.

eg.

$$
\begin{array}{ll}
94 & -6 \\
35 & -65 \\
\hline
29 & /\ 390 \\
\end{array}
$$

= 32 / 90 (carry 3 to the left and add to 29 to get 32)

Case 2 : Integers above the base (100)

Consider the multiplication of 103×105. The technique remains exactly the same and can be written as follows:

```
103     +3
105     +5
_____
108  /  15
```

Notice that here the differences are positive, i.e. + 3 (103 – 100) and + 5 (105 – 100).

The addition in the cross-wise direction gives 108 (103 + 5 or 105 + 3).

The product of the differences is 15 and so the final result is 10815.

Once again, we have to ensure that we have exactly two digits in the second part of the answer, which is achieved by either padding with zeroes or by carry to the left.

eg. Consider the following two cases

a) 101 × 101

```
101     +1
101     +1
_____
102  /  01    (Pad with a zero on the left)
```

b) 113 × 115

```
113     +13              13      +3
115     +15              15      +5
_____          _____
128 /  195               18  /   15
= 12995 (Carry 1 to      = 195  (Carry 1 to
        the left)                the left)
```

Here, 13 and 15 have also been multiplied by using the same technique, where the base has been taken as 10.

This same technique can also be used to find the squares of numbers close to the base. e.g. 96^2 or 104^2.

Case 3 : Integers below and above the base (100)

Consider the multiplication of 97 × 104.

Here 97 is below 100 while 104 is above 100. Using the same technique, we get

$$\begin{array}{r} 97 \quad - \ 3 \\ 104 \quad + \ 4 \\ \hline \end{array}$$

$$101 \ / \ (-12)$$

Here the 2nd part of the answer is −12. Whenever the 2nd part is negative, we carry 1 or more from the left. In this case, we carry 1 from the left which is equivalent to 100 (since the base is 100) and the final answer is obtained as follows :

$$\begin{aligned} 101 \ / \ (-12) \quad &= \quad 100 \ / \ (100 - 12) \\ &= \quad 100 \ / \ 88 = 10088. \end{aligned}$$

This can also be interpreted as $10100 - 12 = 10088$.

Use the technique explained in chapter 1 (all from 9 and last from 10) to subtract from 100.

Once again, the 2nd part of the number should have exactly two digits, since the base has 2 zeros.

Consider the following 2 examples:

a) 101 × 99

$$\begin{array}{r} 101 \quad +1 \\ 99 \quad - \ 1 \\ \hline \end{array}$$

$$\begin{array}{r} 100 \ / \ -1 \\ 99 \ / \ 99 \quad = \ 9999 \end{array}$$

b) 112 × 89

$$
\begin{array}{rr}
112 & +12 \\
89 & -11 \\
\hline
101 ~/~ & (-132)
\end{array}
$$

132 can be obtained by multiplying 12 by 11 as explained in *Chapter 2*. Since 132 is more than 100, we have to carry 2 from the left, which will be equivalent to 200 (2 × 100).

Hence the result would be
$$= 99 ~/~ (200 - 132) = 9968$$

Or as before, we can get the answer by interpreting it as
$$10100 - 132 = 9968.$$

Case 4 : Examples of decimal numbers

Consider 9.8 × 89. We will assume the numbers to be 98 and 89 and carry out the multiplication as shown below :

$$
\begin{array}{rr}
98 & -2 \\
89 & -11 \\
\hline
87 ~/~ & +22
\end{array}
$$

$$= 87 / 22 = 8722.$$

Now, since the original numbers have one digit in the decimal place, we put the decimal point one digit from the right and get the final answer as 872.2.

Case 5 : Examples with bases of 1000 and 10000

Consider the example 997 × 993, where the base is 1000.

$$
\begin{array}{rr}
997 & -3 \\
993 & -7 \\
\hline
990 ~/~ & 021
\end{array}
$$

The procedure remains the same.

The first part of the answer i.e. 990 is obtained by the cross-wise sum, 993 − 3 or 997 − 7.

The second part is obtained by multiplying 3 and 7 and then padding 21 with '0' since we should have 3 digits.

Hence the final answer is 990021.

e.g. Now, consider the example 9998 × 10003, where the base is 10000.

$$
\begin{array}{ll}
9998 & -2 \\
10003 & +3 \\
\hline
10001 \quad / & -0006
\end{array}
$$

The first part of the answer, i.e. 10001 is obtained by the cross-wise sum, 9998 + 3 or 10003 − 2.

The second part is obtained by multiplying −2 and +3 and then padding −6 with '000', since we should have 4 digits (number of zeroes in base 10000). We now carry 1 from left, (equivalent to 10000) and subtract 6 from it to get 9994.

As before, this is 100010000 − 6.

Hence the final answer is 10000/9994 i.e. 100009994.

The method for multiplying decimal numbers has been explained above, wherein we first consider the numbers without decimal points, multiply them using this technique and then place the decimal point at the appropriate location.

Case 6 : Other Bases — 50, 250, 500 etc.

Let us now consider the multiplication of numbers which may be close to other bases like 50, 250, 500 etc. which are multiples of 50. Later, we will see how the method can be extended to bases like 40, which are not multiples of 50. The reader will then be able to extend the method to any base of his choice.

The solved examples are grouped under:
i. Secondary Base of 50
ii. Secondary Base of 250
iii. Secondary Base of 500
iv. Secondary Base of 40 and 60
v. Secondary Base of 300

I. Secondary Base of 50

Example 1 - Consider the multiplication of 42 × 46

Here, we can take the secondary base as 50. The corresponding primary base can be considered as 10 or 100. The scaling factor between the primary and secondary base would be as follows:

Factor = Secondary Base / Primary Base

$$5 = 50 / 10 \quad \text{(If Primary Base = 10)}$$
$$1/2 = 50 / 100 \quad \text{(If Primary Base = 100)}$$

We can use either of these scaling operations and compute the required answer. Let's see each case in detail.

A) Case 1 - Primary base 10 and Secondary base 50 for 42 × 46

- Secondary Base is 50
- Scaling Operation : Factor = 5 = 50 / 10
  (secondary base ÷ primary base)
- One digit to be retained in the right hand part of the answer because primary base has one zero.

The initial part of the method remains the same and is as follows:

Consider : 42 × 46 (Subtract 50 from each number)

$$
\begin{array}{rr}
42 & -8 \\
46 & -4 \\
\hline
38 & / \quad 32 \\
\hline
\end{array}
$$

Now, we have to scale up the left part of this intermediate result by multiplying by 5.

Since 38 × 5 = 190, the answer becomes 190 / 32.

Since the primary base (10) contains one zero, we will retain one digit in the right part of the answer and carry 3 to the left.

So, the final answer becomes 1932.

B) Case 2 - Primary base 100 and Secondary base 50 for 42 × 46

● Secondary Base is 50
● Scaling Operation : Factor = 1/2 = 50 / 100
                         (secondary base ÷ primary base)
● Two digits to be retained in the right hand part of the answer.

The technique remains the same and is as follows:

Consider : 42 × 46 (Subtract 50 from each number)

$$
\begin{array}{rr}
42 & -8 \\
46 & -4 \\
\hline
38 & / \quad 32 \\
\hline
\end{array}
$$

Now, we have to scale down the left part of this intermediate result by multiplying it by 1/2.

Since 38 / 2 = 19, the answer becomes **19 / 32** = 1932.

Since the primary base is 100, which contains two zeros, we will retain two digits in the right part of the answer.

So, the final answer becomes 1932.

Wow, isn't that wonderful? Both the operations give us the same result.

Specific Situations

Let us now consider a few specific situations :
- A fraction is obtained during the scaling operation, and/or
- The right part is negative

We will take an example having both these situations.

Example 2 - Consider the multiplication of 52 × 47

C) Case 1 - Primary base 10, Secondary base 50 for 52 × 47

- Secondary Base is 50
- Scaling Operation : Factor = 5 = 50 / 10
                        (secondary base ÷ primary base)
- One digit to be retained in the right hand part of the answer

Consider : 52 × 47 (Subtract 50 from each number)

$$
\begin{array}{rr}
52 & +2 \\
47 & -3 \\
\hline
49 & / (-6)
\end{array}
$$

We will scale up this intermediate result by multiplying 49 by 5.

Since 49 × 5 = 245, the answer becomes **245 / (–6)**.

Since the primary base is 10, we will carry one from the left which is equal to 10 and subtract 6 from it.

So the final answer becomes 244/4 = 2444.

This can also be considered as 2450 − 6 = 2444.

D) Case 2 - Primary base 100 and Secondary base 50 for 52 × 47

- Secondary Base is 50
- Scaling Operation : Factor = 1/2 = 50 / 100
  (secondary base ÷ primary base)
- Two digits to be retained in the right hand part of the answer

Consider : 52 × 47 (Subtract 50 from each number)

$$
\begin{array}{ll}
52 & +2 \\
47 & -3 \\
\hline
49 & / (-6) \\
\hline
\end{array}
$$

We will scale down this intermediate result by dividing 49 by 2.

Since 49 / 2 = 24.5, the answer becomes 24½ / (−6).

Since the primary base is 100, we will carry 1/2 from the left. This is equal to a carry of 50 (primary base 100/2). We will subtract 6 from 50 giving 44.

So the final answer becomes 24/44 = 2444.

Since 24.5 hundreds is 2450, this can be interpreted also as

2450 / (−6) = 2444

We get the same answer irrespective of the primary base. The fractional part can be avoided if we choose a scaling operation involving multiplication instead of division.

II. Secondary Base of 250

Example 2 - Consider the multiplication of 266 × 235

E) Primary base 1000 and Secondary base 250 for 235 × 266

- Secondary Base is 250
- Scaling Operation : Factor = 1/4 = 250 / 1000
                          (secondary base ÷ primary base)
- Three digits to be retained in the right hand part of the answer

Consider : 235 × 266 (Subtract 250 from each number)

Consider the steps

$$
\begin{array}{ll}
235 & -15 \\
266 & +16 \\
\hline
251 & /\ (-240)
\end{array}
$$

We will scale down this intermediate result by dividing 251 by 4.

Since 251 / 4 = 62¾, the answer becomes 62¾ / (−240).

Since the primary base is 1000, we will carry 3/4 from the left, which is equal to a carry of 750 (primary base 1000 * ¾) and subtract 240 from 750 giving 510.

So the final answer becomes 62/510 = 62510.

Or, it can be considered as 62750 − 240 = 62510.

III. Secondary Base of 500

Example 2 - Consider the multiplication of 532 × 485

F) Primary base 100 and Secondary base 500 for 532 × 485

- Secondary Base is 500
- Scaling Operation : Factor = 5 = 500 / 100
  (secondary base ÷ primary base)
- Two digits to be retained in the right hand part of the answer

Consider : 532 × 485 (Subtract 500 from each number)

Consider the steps

$$
\begin{array}{ll}
532 & +32 \\
485 & -15 \\
\hline
517 & / \;(-480)
\end{array}
$$

We will scale up this intermediate result by multiplying 517 by 5.

Since 517 × 5 = 2585, the answer becomes 2585 / (−480).

Since the primary base is 100, we will carry 5 from the left, which is equal to a carry of 500 (primary base 100 * 5) and subtract 480 from 500 giving 20.

So the final answer becomes 2580/020 = 258020.

Or, it can be considered as 258500 − 480 giving 258020.

G) Primary base 1000 and Secondary base 500 for 532 × 485

- Secondary Base is 500
- Scaling Operation : Factor = 1/2 = 500 / 1000
  (secondary base ÷ primary base)

● Three digits to be retained in the right hand part of the answer

Consider the steps

$$
\begin{array}{ll}
532 & +32 \\
485 & -15 \\
\hline
517 & / \quad (-480)
\end{array}
$$

We will scale down this intermediate result by dividing 517 by 2.

Since $517 / 2 = 258½$, the answer becomes $258½ / (-480)$.

Since the primary base is 1000, we will carry 1/2 from the left, which is equal to a carry of 500 (primary base 1000 * 1/2) and subtract 480 from 500 giving 20.

So the final answer becomes $258/020 = 258020$.

The left part of the answer is 258.5 hundreds which is equal to 258500.

So, it can be considered as $258500 - 480$ giving 258020.

IV. Secondary Bases like 40, 60

Example 3 - Consider the multiplication of 32 × 48

H) Primary base 10 and Secondary base 40 for 32 x 48

● Secondary Base = 40
● Scaling Operation : Factor = 4 = 40 / 10
     (secondary base ÷ primary base)
● One digit to be retained in the right hand part of the answer

Consider : 32 × 48 (Subtract 40 from each number)

$$32 \qquad -8$$
$$48 \qquad +8$$
$$\overline{}$$
$$40 \quad / \quad (-64)$$
$$\overline{}$$

We will scale up this intermediate result by multiplying 40 by 4.

Since $40 \times 4 = 160$, the answer becomes $160 / (-64)$.

Since the primary base is 10, we will carry 7 from the left, which is equal to a carry of 70 (primary base 10 * 7) and subtract 64 from 70 giving 6.

This is also equivalent to $1600 - 64$.

So, the final answer is 1536.

Example 4 - Consider the multiplication of 64 × 57

I) Primary base 10 and Secondary base 60 for 64 × 57

- Secondary Base = 60
- Scaling Operation : Factor = 6 = 60 / 10
                           (secondary base ÷ primary base)
- One digit to be retained in the right hand part of the answer

Consider : 64 × 57 (Subtract 60 from each number)

$$64 \qquad +4$$
$$57 \qquad -3$$
$$\overline{}$$
$$61 \quad / \quad (-12)$$
$$\overline{}$$

We will scale up this intermediate result by multiplying 61 by 6.

Since $61 \times 6 = 366$, the answer becomes $366 / (-12) = 364 / 8$.

Since the primary base is 10, we will carry 2 from the left, which is equal to a carry of 20 (primary base 10 * 2) and subtract 12 from 20 giving 8.

This is also equivalent to $3660 - 12$.

So, the final answer is 3648.

V. Secondary Base of 300

Example 5 - Consider the multiplication of 285 × 315

J) Primary base 100 and Secondary base 300 for 285 × 315

- Secondary Base is 300
- Scaling Operation : Factor = 3 = 300 / 100
  (secondary base ÷ primary base)
- Two digits to be retained in the right hand part of the answer

Consider : 285 × 315 (Subtract 300 from each number)

$$
\begin{array}{ll}
285 & -15 \\
315 & +15 \\
\hline
300 & / \quad (-225)
\end{array}
$$

We will scale up this intermediate result by multiplying 300 by 3.

Since $300 \times 3 = 900$, the answer becomes $900 / (-225)$.

Since the primary base is 100, we will carry 3 from the left which is equal to a carry of 300 (primary base 100 * 3) and subtract 225 from 300 giving 75.

So the final answer becomes $897/75 = 89775$.

Conclusion

This section gives an idea of how a variety of combinations of primary and secondary bases can be used to arrive at the final answer quickly. The reader should choose the secondary base carefully which can reduce the time for computation drastically.

5. *Examples for multiplication of numbers below the base (100)*

Sr No	Exercise	Answer
1.	97 × 83	8051
2.	96 × 98	9408
3.	97 × 92	8924
4.	88 × 98	8624
5.	87 × 840	73080
6.	97 × 93	9021
7.	88 × 89	7832
8.	85 × 8.9	756.50
9.	97 × 320	31040
10.	9.8 × 0.95	9.31
11.	8.4 × 0.089	0.7476
12.	8.8 × 9.7	85.36
13.	7.6 × 8.7	66.12
14.	0.0092 × 0.0092	0.00008464
15.	8.7 × 93	809.10
16.	0.77 × 98	75.46
17.	86 × 0.97	83.42
18.	9.6 × 79	758.40
19.	0.085 × 93	7.905
20.	6.9 × 8.8	60.92

6. Examples for multiplication of numbers above the base (100)

Sr No	Exercise	Answer
1.	104 × 105	10920
2.	103 × 108	11124
3.	1110 × 107	118770
4.	101 × 121	12221
5.	143 × 104	14872
6.	115 × 113	12995
7.	112 × 1070	119840
8.	113 × 103	11639
9.	10.2 × 10.5	107.10
10.	10.1 × 105	1060.50
11.	1.05 × 10.2	10.71
12.	0.115 × 10.3	1.1845
13.	1.06 × 10.4	11.024
14.	11.5 × 1.01	11.615
15.	0.112 × 1030	115.36
16.	0.104 × 107	11.128
17.	119 × 131	15589
18.	1.12 × 10.3	11.536
19.	0.143 × 109	15.587
20.	107 × 11.2	1198.40

7. Examples for multiplication of numbers above and below the base (100)

Sr No	Exercise	Answer
1.	94 × 108	10152
2.	96 × 109	10464
3.	920 × 106	97520
4.	88 × 112	9856
5.	91 × 107	9737
6.	76 × 1090	82840
7.	9.3 × 1.15	10.695
8.	95 × 111	10545
9.	10.2 × 9.8	99.96
10.	10.6 × 0.97	10.282
11.	108 × 0.97	104.76
12.	0.0095 × 1.06	0.01007
13.	0.94 × 1.12	1.0528
14.	0.88 × 10.6	9.328
15.	8.7 × 13.5	117.45
16.	95 × 106	10070
17.	9.4 × 11.2	105.28
18.	10.5 × 9.7	101.85
19.	0.0085 × 10.4	0.0884
20.	0.79 × 10.7	8.453

8. Examples for squares of numbers below the base (100)

Sr No	Exercise	Answer
1.	96^2	9216
2.	98^2	9604
3.	87^2	7569
4.	9.8^2	96.04

5.	0.87^2	0.7569
6.	89^2	7921
7.	94^2	8836
8.	0.99^2	0.9801
9.	0.097^2	0.009409
10.	880^2	774400
11.	0.092^2	0.008464
12.	970^2	940900
13.	9.3^2	86.49
14.	0.86^2	0.7396
15.	9200^2	84640000
16.	9.6^2	92.16
17.	0.89^2	0.7921
18.	88^2	7744
19.	0.97^2	0.009409
20.	0.77^2	0.5929

9. Examples for squares of numbers above the base (100)

Sr No	Exercise	Answer
1.	102^2	10404
2.	105^2	11025
3.	104^2	10816
4.	112^2	12544
5.	109^2	11881
6.	10.6^2	112.36
7.	1.08^2	1.1664
8.	11.3^2	127.69
9.	12.5^2	156.25
10.	1130^2	1276900
11.	0.0109^2	0.00011881

12.	111^2	12321
13.	13.5^2	182.25
14.	0.113^2	0.012769
15.	11400^2	129960000
16.	117^2	13689
17.	11.9^2	141.61
18.	12.4^2	153.76
19.	0.119^2	0.014161
20.	1.32^2	1.7424

10. Examples for multiplication of numbers (3-digits) below the base (1000)

Sr No	Exercise	Answer
1.	899 × 995	894505
2.	994 × 885	879690
3.	986 × 993	979098
4.	777 × 995	773115
5.	386 × 998	385228
6.	975 × 882	859950
7.	987 × 992	979104
8.	991 × 355	351805
9.	113 × 991	111983
10.	994 × 995	989030
11.	92.7 × 9.95	922.365
12.	93.4 × 9.97	931.198
13.	0.0988 × 978	96.6264
14.	9.72 × 99.6	968.112
15.	99.9 × 9.53	952.047
16.	98.1 × 0.988	96.9228

17.	9.72 × 99.6	968.112
18.	0.0992 × 0.194	0.0192448
19.	9.92 × 9.97	98.9024
20.	9.93 × 235	2333.55

11. Examples for multiplication of numbers (3-digits) above the base (1000)

Sr No	Exercise	Answer
1.	1006 × 1005	1011030
2.	1044 × 1004	1048176
3.	1010 × 1013	1023130
4.	1014 × 1019	1033266
5.	1013 × 10030	10160390
6.	1009 × 1006	1015054
7.	1.005 × 1002	1007.010
8.	1015 × 1012	1027180
9.	1.022 × 100.1	102.3022
10.	1.014 × 10.03	10.17042
11.	1018 × 1004	1022072
12.	10.15 × 1010	10251.50
13.	1011 × 1023	1034253
14.	1031 × 1007	1038217
15.	1055 × 1.005	1060.275
16.	10.06 × 1012	10180.72
17.	1008 × 10.09	10170.72
18.	1034 × 1002	1036068
19.	10.21 × 1010	10312.10
20.	1054 × 1002	1056108

12. Examples for multiplication of numbers (3-digits) above & below the base (1000)

Sr No	Exercise	Answer
1.	1004 × 998	1001992
2.	1008 × 990	997920
3.	1007 × 995	1001965
4.	998 × 1004	1001992
5.	997 × 1001	997997
6.	985 × 1100	1083500
7.	1112 × 998	1109776
8.	1010 × 976	985760
9.	1004 × 999	1002996
10.	1.010 × 99.5	100.495
11.	0.1012 × 0.983	.0994796
12.	1.015 × 98.9	100.3835
13.	1.022 × 9.92	10.13824
14.	102.5 × 8.88	910.20
15.	1.021 × 0.0997	0.1017937
16.	1.019 × 98.7	100.5753
17.	1.007 × 0.996	0.1002972
18.	103.2 × 8.86	914.352
19.	1011 × 986	996846
20.	10.13 × 991	10.03883

13. Examples for Squares of numbers (3-digits) below the base (1000)

Sr No	Exercise	Answer
1.	996^2	992016
2.	99.4^2	9880.36
3.	9.98^2	99.6004

4.	0.993^2	0.986049
5.	888^2	788544
6.	9950^2	99002500
7.	997^2	994009
8.	9.99^2	99.8001
9.	0.985^2	0.970225
10.	98.4^2	9682.56
11.	0.0997^2	0.00994009
12.	9040^2	81721600
13.	9.94^2	98.8036
14.	88.9^2	7903.21
15.	0.992^2	0.984064
16.	9.92^2	98.4064
17.	98.1^2	9623.61
18.	88.7^2	7867.69
19.	786^2	972196
20.	0.988^2	0.976144

14. Examples for Squares of numbers (3-digits) above the base (1000)

Sr No	Exercise	Answer
1.	1004^2	1008016
2.	1008^2	1016064
3.	1012^2	1024144
4.	10.18^2	103.6324
5.	100.5^2	10100.25
6.	1.007^2	1.014049
7.	11.22^2	125.8884
8.	101.4^2	10281.96
9.	10.09^2	101.8081

10.	100.9^2	10180.81
11.	104.4^2	10899.36
12.	1.067^2	1.138489
13.	1.015^2	1.030225
14.	10.22^2	104.4484
15.	0.1014^2	0.01028196
16.	10.26^2	105.2676
17.	100.8^2	10160.64
18.	1032^2	1065024
19.	1106^2	1223236
20.	1.0172^2	1.034289

15. Examples for multiplication of numbers (4-digits) below the base (10000)

Sr No	Exercise	Answer
1.	9995×9992	99870040
2.	9873×9952	98256096
3.	9788×9998	97860424
4.	8899×9988	88883212
5.	9992×9878	98700976
6.	1535×9989	15333115
7.	98.72×9.992	986.41024
8.	0.9925×9.921	9.8465925
9.	9.972×88.89	886.41108
10.	99.75×899.9	89765.025
11.	9898×9994	98920612
12.	8989×9899	88982111
13.	0.8898×9962	8864.1876
14.	9988×9977	99650276

15.	9.897 × 9999	98960.103
16.	1499 × 9986	14969014
17.	9.692 × 899.9	8721.8308
18.	0.1819 × 9926	1805.5394
19.	8793 × 99.72	876837.96
20.	0.1697 × 9959	1690.0423

16. Examples for multiplication of numbers (4-digits) above the base (10000)

Sr No	Exercise	Answer
1.	10002 × 10008	100100016
2.	10015 × 10020	100350300
3.	12154 × 10003	121576462
4.	10004 × 10010	100140040
5.	100.08 × 10185	1019314.80
6.	108.49 × 10.007	1085.65943
7.	10.013 × 1001.5	10028.0195
8.	1.0083 × 101.10	101.939130
9.	11572 × 10004	115766288
10.	12.728 × 10006	127356.368
11.	103.42 × 100.11	10353.3762
12.	11022 × 1100.4	12128608.80
13.	10.025 × 100.24	1004.90600
14.	1000.7 × 134.44	134534.108
15.	1.0012 × 101.35	101.471620
16.	10012 × 10006	100180072
17.	10.051 × 10011	100620.561
18.	1.0017 × 101.36	101.532312
19.	12.672 × 10004	126770.688
20.	11472 × 101.09	1159704.48

17. Examples for Squares of numbers (4-digits) below the base (10000)

Sr No	Exercise	Answer
1.	9996^2	99920016
2.	9994^2	99880036
3.	9988^2	99760144
4.	9993^2	99860049
5.	988.8^2	977725.44
6.	9.998^2	99.960004
7.	98.72^2	9745.6384
8.	9750^2	95062500
9.	994.5^2	989030.25
10.	999.9^2	999800.01
11.	9978^2	99560484
12.	98.99^2	9799.0201
13.	9950^2	99002500
14.	984.5^2	969240.25
15.	988.9^2	977923.21
16.	9899^2	97990201
17.	9878^2	97574884
18.	9.993^2	99.860049
19.	9969^2	99380961
20.	987.9^2	975946.41

18. Examples for Squares of numbers (4-digits) above the base (10000)

Sr No	Exercise	Answer
1.	10002^2	100040004
2.	100.04^2	10008.0016

3.	10006^2	100120036
4.	10.015^2	100.300225
5.	1.0007^2	1.00140049
6.	100.13^2	10026.0169
7.	10025^2	100500625
8.	1001.4^2	1002801.96
9.	1002.2^2	1004404.84
10.	1001.9^2	1003803.61
11.	10022^2	100440484
12.	10015^2	100300225
13.	1006.4^2	1012840.96
14.	10.051^2	101.022601
15.	100.07^2	10014.0049
16.	10023^2	100460529
17.	10129^2	102596641
18.	10.008^2	100.160064
19.	1.0029^2	1.00480841
20.	101.04^2	102090816

19. Examples for multiplication of Mixed Numbers

Sr No	Exercise	Answer
1.	38 × 62	2356
2.	73 × 84	6132
3.	583 × 603	351549
4.	224 × 252	56448
5.	129 × 251	32379
6.	485 × 531	257535
7.	271 × 502	136042

8.	162 × 173	28026
9.	18.9 × 20.7	391.23
10.	83 × 198	16434
11.	5.95 × 62.5	371.875
12.	88.2 × 0.921	81.2322
13.	326 × 3.02	984.52
14.	0.785 × 81.5	63.9775
15.	74.1 × 7.52	557.232
16.	7.7 × 0.86	6.622
17.	386 × 251	96886
18.	19.8 × 30.5	603.90
19.	5.96 × 64.5	384.42
20.	76.6 × 81.7	6258.22

CHAPTER 5

Multiplication (Urdhva Tiryak)

Vertically and Crosswise

We will now consider the other method of multiplication viz. the 'Urdhva Tiryak' or cross-multiplication technique. It consists of vertical and cross multiplication between the various digits and gives the final result in a single line.

We will consider the multiplication of the following :

- Two 2-digit numbers
- Two 3-digit numbers
- Two 4-digit numbers
- 2-digit with 3-digit numbers
- 2-digit with 4-digit numbers
- 3-digit with 4-digit numbers

Once the technique is clear, it can be extended to the multiplication of any number of digits.

Case 1 : 2-digit multiplication

Consider the following multiplication as done by the traditional method.

$$
\begin{array}{r}
3\,8 \\
\times \quad 5\,2 \\
\hline
1\,9\,0\,0 \\
+ \quad\quad 7\,6 \\
\hline
1\,9\,7\,6
\end{array}
$$

The **Vedic Maths** method consists of computing three digits a, b and c as shown in the diagram

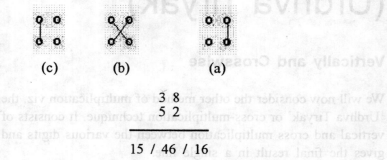

(c)   (b)   (a)

$$
\begin{array}{r}
3\,8 \\
5\,2 \\
\hline
15 \ / \ 46 \ / \ 16 \\
\hline
\end{array}
$$

Steps :

- **Digit (a) :** Multiply the right-most digits 8 and 2 to get 16.

- **Digit (b) :** Carry out a cross multiplication between 3 and 2, and 8 and 5 to get 6 and 40 respectively and on adding them, we get 46.

- **Digit (c) :** Multiply the left-most digits 3 and 5 to get 15.

We have to retain one digit at each of the three locations and carry the surplus to the left.

Hence in

$$15 \ / \ 46 \ / \ 16$$

we start from the right, retain 6 and carry 1 to the left. This gives 47 in the middle location.

Now, we retain 7 and carry 4 to the left. We add 4 to 15 to get 19.

This gives 19 / 7 / 6, i.e. 1976 which is the required answer.

To speed up this operation, the carry digit can be written just below the number on the left. It can then be added to the number to its left at each operation itself. Let's see the multiplication again.

$$
\begin{array}{c}
3\ 8 \\
5\ 2 \\
\hline
19 / 7 / 6 \\
4\ \ \ \ 1
\end{array}
$$

Here, the right-most vertical multiplication is carried out first to get $8 \times 2 = 16$. We write 6 and write the carry 1 as shown.

Next we carry out the cross multiplication $3 \times 2 + 5 \times 8 = 46$, add the carry (1) to it and get 47. We write down 7 and the carry digit 4 as shown.

The 3rd product is 3×5 which gives 15 to which we add the carry digit of 4 and get 19.

As you can see, the method gives you the final answer faster in just one line.

Let's try another example.

$$
\begin{array}{r}
7\ \ 6 \\
\times \quad 4\ \ 3 \\
\hline
32 :\ 6 :\ 8 \\
4 \quad\ \ 1
\end{array}
$$

Steps :

- **Digit (a) :** Multiply the right-most digits 6 and 3 to get **18**. We write 8 and show 1 as the carry on its left.

- **Digit (b)** : Carry out a cross multiplication between 7 and 3 and 6 and 4 to get 21 and 24 respectively and add them to get 45. Add the carry digit 1 and get **46**. Write down 6 and show the carry as 4.

- **Digit (c)** : Multiply the left-most digits 7 and 4 to get 28. Add the carry digit 4 to it to get **32**.

20. Examples for 2-digit cross multiplication

Sr No	Exercise	Answer
1.	24 × 83	1992
2.	42 × 54	2268
3.	66 × 86	5676
4.	27 × 78	2106
5.	870 × 45	39150
6.	4.8 × 0.54	2.592
7.	33 × 0.0065	0.2145
8.	74 × 0.89	65.86
9.	97 × 38	3686
10.	44 × 620	27280
11.	38 × 4300	163400
12.	2.8 × 3.5	9.80
13.	4.8 × 5.3	25.44
14.	6.8 × 7.2	48.96
15.	73 × 8.5	620.50
16.	980 × 3.4	3332
17.	0.0088 × 0.036	0.0003168
18.	4.6 × 0.0033	0.01518
19.	53000 × 620	32860000
20.	0.081 × 3.2	0.2592

Case 2 : 3-digit multiplication

Let us now consider multiplication of two 3-digit numbers.

Consider the following example

```
        3   2   6
  ×     2   5   8
  ─────────────────
    6   5   2   0   0
  + 1   6   3   0   0
  +     2   6   0   8
  ─────────────────
    8   4   1   0   8
```

The **Vedic Maths** method consists of computing five digits a, b, c, d, e as shown in the diagram :

(e)   (d)   (c)   (b)   (a)

We will use the method explained in the previous case in which the carry digit at each step was added to the number to its left.

Steps :

*   **Digit (a) :** Multiply the right-most digits 6 and 8 to get **48**. We write 8 and show 4 as the carry, on its left side.

*   **Digit (b) :** Carry out a 2-digit cross multiplication between 2 and 8 (16), and 6 and 5 (30).
    *   Add the products 16 and 30 to get 46.
    *   Add the carry digit 4 and get **50**.
    *   Write down 0 and show the carry as 5.

*   **Digit (c) :** Carry out a 3-digit cross multiplication between 3 and 8 (24), and 6 and 2 (12) and 2 and 5 (10).

- • Add the three products 24, 12 and 10 to get 46.
- • Add the carry digit 5 and get **51**.
- • Write down 1 and show the carry as 5.

- • **Digit (d)** : Carry out a 2-digit cross multiplication between 3 and 5 (15) and 2 and 2 (4).
  - • Add 15 and 4 to get 19.
  - • Add the carry digit 5 to get **24,**
  - • Write down 4 and show 2 as the carry.

- • **Digit (e)** : Multiply the left-most digits 3 and 2, to get 6. Add the carry digit 2 to it to get **8**.

Finally, the multiplication looks as follows :

$$
\begin{array}{r}
3 \quad 2 \quad 6 \\
\times \quad 2 \quad 5 \quad 8 \\
\hline
8 : 4 : 1 : 0 : 8 \\
2 \quad\; 5 \quad\; 5 \quad\; 4
\end{array}
$$

Let's try another example, the computation is as shown.

$$
\begin{array}{r}
4 \quad 5 \quad 3 \\
\times \quad 6 \quad 2 \quad 1 \\
\hline
28 : 1 : 3 : 1 : 3 \\
4 \quad\; 3 \quad\; 1
\end{array}
$$

Case 3 : *Multiplying 3-digit and 2-digit numbers*

If we have to multiply a 3-digit with a 2-digit number, we can treat the 2-digit number as a 3-digit number by padding it with a zero on the left.

Eg. 354 × 45 can be written as 354 × 045 and the same technique can be used.

21. Examples for 3-digit cross multiplication

Sr No	Exercise	Answer
1.	752 × 303	227856
2.	189 × 215	40635
3.	32.5 × 43.6	1417
4.	1240 × 801	993240
5.	775 × 576	446400
6.	7120 × 41500	295480000
7.	0.633 × 8.22	5.20326
8.	42.3 × 536	22672.80
9.	7.18 × 23.3	167.294
10.	8.99 × 121	1087.79
11.	21300 × 1.45	30885
12.	0.0325 × 6.12	0.1989
13.	0.00312 × 0.235	0.0007332
14.	6.34 × 0.318	2.01612
15.	7.15 × 0.000323	0.00230945
16.	23100 × 3.14	72534
17.	802 × 316	253432
18.	910 × 861	783510
19.	452 × 0.00728	3.29056
20.	90500 × 0.623	56381.50

22. More Examples - 3 digits × 2 digits

Sr No	Exercise	Answer
1.	81 × 233	18873
2.	720 × 29	20880

3.	121 × 25	3025
4.	42 × 32.2	1352.40
5.	175 × 0.35	61.25
6.	143 × 6.8	972.40
7.	16.8 × 0.67	11.256
8.	71 × 2.32	164.72
9.	625 × 0.016	10.00
10.	63 × 43.3	2727.90
11.	215 × 0.53	113.95
12.	86 × 163	14018
13.	1460 × 1.6	2336
14.	723 × 0.012	8.676
15.	340 × 0.126	42.84
16.	12500 × 0.54	6750
17.	332 × 0.0042	1.3944
18.	2180 × 5.8	12644
19.	253 × 4820	1219460
20.	784 × 21200	16620800

Case 4 : 4-digit multiplication

The techniques discussed till now can be extended further for four digits. The diagram below can be used.

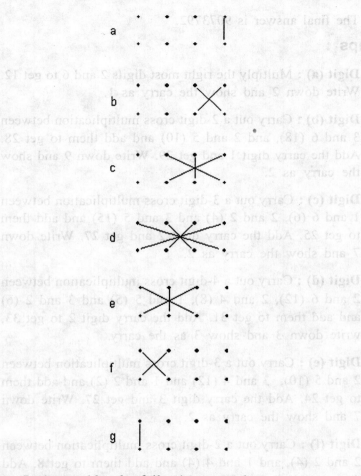

Let's consider the following example

```
      2  1  3  2
   ×  4  2  5  6
   ─────────────────────────────────
   8 : 8 : 24 : 31 : 25 : 28 : 12
   ─────────────────────────────────
   9 : 0 : 7 :  3 :  7 :  9 :  2
      1    2    3    2    2    1
```

The final answer is 9073792.

Steps :

- **Digit (a)** : Multiply the right most digits 2 and 6 to get 12. Write down 2 and show the carry as 1.

- **Digit (b)** : Carry out a 2-digit cross multiplication between 3 and 6 (18), and 2 and 5 (10) and add them to get 28. Add the carry digit 1 and get 29. Write down 9 and show the carry as 2.

- **Digit (c)** : Carry out a 3-digit cross multiplication between 1 and 6 (6), 2 and 2 (4) and 3 and 5 (15) and add them to get 25. Add the carry digit 2 and get **27**. Write down 7 and show the carry as 2.

- **Digit (d)** : Carry out a 4-digit cross multiplication between 2 and 6 (12), 2 and 4 (8), 1 and 5 (5) and 3 and 2 (6) and add them to get 31. Add the carry digit 2 to get **33**, write down 3 and show 3 as the carry.

- **Digit (e)** : Carry out a 3-digit cross multiplication between 2 and 5 (10), 3 and 4 (12) and 1 and 2 (2) and add them to get 24. Add the carry digit 3 and get **27**. Write down 7 and show the carry as 2.

- **Digit (f)** : Carry out a 2-digit cross multiplication between 2 and 2 (4), and 1 and 4 (4) and add them to get **8**. Add the carry digit 2 and get 10. Write down 0 and show the carry as 1.

- **Digit (g)** : Multiply the left most digits 2 and 4 to get 8. Add the carry digit 1 to it to get **9**.

Let's try another example, the computation is as shown:

$$
\begin{array}{r}
5453 \\
\times \quad 3621 \\
\end{array}
$$

15	:	42	:	49	:	52	:	32	:	11	:	3

19	7	4	5	3	1	3
	4	5	5	3	1	

The final result is 19745313.

Case 5 : Multiplying 4-digit and 3-digit numbers

If we have to multiply a 4-digit with a 3-digit number, we can treat the 3-digit number as a 4-digit number by padding it with a zero on the left.

Eg. 3542 × 453 can be written as 3542 × 0453 and the same technique can be used.

23. Examples for 4-digit cross multiplication

Sr No	Exercise	Answer
1.	3125 × 7218	22556250
2.	1284 × 4015	5155260
3.	2115 × 3004	6353460
4.	2222 × 3333	7405926
5.	1005 × 4128	4148640
6.	2010 × 314	631140
7.	28.14 × 7.135	200.7789
8.	13500 × 783	10570500
9.	21920 × 0.1342	2941.664
10.	3780 × 1992	7529760
11.	16654 × 2323	38687242
12.	43244 × 2419	104607236

13.	56032 × 0.5164	28934.9248
14.	12988 × 3.214	41743.432
15.	32176 × 2163	69596688
16.	2152 × 5213	11218376
17.	1020 × 4601	4693020
18.	39.42 × 1992	39104.64
19.	1651 × 32.09	52980.59
20.	4603 × 1876	8635228

24. Examples for multiplication - 4-digits with 3-digits

Sr No	Exercise	Answer
1.	3125 × 721	2253125
2.	128 × 5012	641536
3.	2110 × 30.4	64144
4.	2432 × 444	1079808
5.	1876 × 4.12	7729.12
6.	2124 × 3.14	6669.36
7.	28.14 × 7.35	206.829
8.	13520 × 7.83	105861.60
9.	21920 × 0.134	2937.28
10.	37880 × 199	7538120
11.	34208 × 321	10980768
12.	5432 × 7.32	39762.24
13.	23098 × 356	8222888
14.	8743 × 12.9	112784.70
15.	2317 × 567	1313739
16.	12.4 × 2608	32339.20
17.	4015 × 213	855195
18.	2119 × 320	678080
19.	9.19 × 6013	55259.47
20.	3.514 × 0.154	0.541156

CHAPTER 6

Division

The operation of division is basically the reverse of multiplication. In the traditional approach, it is a slow and tedious method with a considerable amount of guesswork. At each step, an approximate quotient digit is guessed, which is then multiplied by the divisor and the remainder is computed. If the remainder is negative or contains additional multiples of the divisor, the quotient digit is recomputed. As the number of digits in the divisor increase, the division becomes progressively slower.

Vedic maths provides a very elegant method wherein the division is reduced to a division by a single digit in most cases or at most to two small digits like 12, 13 etc in a few cases. At each step, the division by a 1-digit or 2-digit divisor is followed by a multiplication, based on the 'Urdhav Tiryak' approach and finally a subtraction.

The reduction in both time and mental strain using the Vedic Maths technique is astounding.

Consider the example of dividing 87108 by 84 which is a 2-digit divisor. The traditional method is shown below :

$$
\begin{array}{r|r|r}
84 & 87108 & 1037 \\
 & -\ 84 & \\
\hline
 & 31 & \\
 & -\ \ 0 & \\
\hline
 & 310 & \\
 & -\ 252 & \\
\hline
 & 588 & \\
 & -\ 588 & \\
\hline
 & 0 & \\
\end{array}
$$

At each stage, we have to mentally carry out a trial division and then write down the quotient which gives either a positive or zero remainder. This is a slow and tedious procedure and hence the operation of division is detested. As the number of digits increase in the divisor, the process becomes more and more slow and time consuming.

The vedic method has been devised beautifully in which the divisions have been reduced to a one-digit division. This speeds up the process considerably and makes it an enjoyable experience.

Case 1 : Division by a number with a flag of one digit (no remainder)

Let's consider the same example again and carry out division by the vedic method.

Divide 87108 by 84

We will write the divisor as 8^4 where 4 is called the flag. The division will be carried out by a single digit viz 8.

```
  4  |
  8  |   8     7     1     0  :  8
     |   0     3     7     2
     | - 4   - 0   - 12   –28
     | _____
     |   3    31    58     0
     |   1     0     3     7  :  0
```

Steps :

- Place a colon one digit from the right (= number of digits of flag)

- Divide 8 by 8 to get quotient as 1 and remainder as 0

- Carry the remainder 0 to the next digit 7 and get **7 (called Gross dividend, GD)**

- Multiply the quotient 1 by the flag 4 to get 4 and subtract it from 7 to get **3 (called Net dividend, ND)**

- Divide 3 by 8 to get the quotient as 0 and remainder as 3

- Carry the remainder 3 to the next digit 1 and get **31 (GD)**

- Multiply the quotient 0 by the flag 4 to get 0 and subtract it from 31 to get **31 (ND)**

- Divide 31 by 8 to get the quotient as 3 and remainder as 7

- Carry the remainder 7 to the next digit 0 and get 70.

- Multiply the quotient 3 by the flag 4 to get 12 and subtract it from 70 to get **58**

- Divide 58 by 8 to get the quotient as 7 and remainder as 2

- At this point, the integer part of the final quotient has been obtained since we have reached the colon. All the balance digits would be the decimal part

- Carry the remainder 2 to the next digit 8 and get **28**

- Multiply the quotient 7 by the flag 4 to get 28 and subtract it from 28 to get **0**

- Since the difference is now 0, the division is completed and we get the answer as **1037.**

Important points :

- At each stage, the net dividend is obtained by multiplying the quotient in the previous column by the flag digit

- The net dividend was always positive.

- Later, we will see what adjustments have to be made if the net dividend becomes negative in any column.

Second Example

Divide 43409 by 83

```
       3
 8  |  43    4     0  :  9
    |        3     3     0
    |      - 15   -6    -9
    |      ─────────────────
    |       19    24     0
    |─────────────────────
    |        5     2     3  :  0
```

Steps :

- Place a colon one digit from the right (= number of digits of flag)

- Divide 43 by 8 to get quotient as 5 and remainder as 3

- Carry the remainder 3 to the next digit 4 and get **34**

- Multiply the quotient 5 by the flag 3 to get 15 and subtract it from 34 to get **19**

- Divide 19 by 8 to get the quotient as 2 and remainder as 3

- Carry the remainder 3 to the next digit 0 and get **30**

- Multiply the quotient 2 by the flag 3 to get 6 and subtract it from 30 to get **24**

- Divide 24 by 8 to get the quotient as 3 and remainder as 0

- Carry the remainder 0 to the next digit 9 and get **09**. At this point, the integer part of the final quotient has been obtained since we have reached the colon. All the balance digits would be the decimal part.

- Multiply the quotient 3 by the flag 3 to get 9 and subtract it from 9 to get **0**

- Since the difference is now 0, the division is completed and we get the answer as **523**.

   Try   a) 43412 / 83   and   b) 43403 / 73

Case 2 : Division by a number with a flag of one digit (with remainder)

Consider the division of 26382 by 77

7							
7	26	3	8 :	2	0	0	0
		5	4	6	6	4	5
		- 21	- 28	- 14	- 42	-14	-21
		32	20	48	18	26	29
	3	4	2 :	6	2	3	

The quotient is 342 and the remainder is 48, which appears under the column just after the colon. The final answer in decimal form is 342.623.

Steps :

*   The steps as explained above are carried out till we reach the column after the colon which shows the remainder as 48. If we want to get the answer in decimal form, we have to carry out division further by using the same procedure.

*   Divide 48 by 7 to get quotient as 6 and remainder as 6

*   Now, we attach a zero to the dividend and write the remainder to its left, giving as the next number as 60.

*   Multiply the quotient 6 by the flag 7 to get 42 and subtract it from 60 to get 18

*   Divide 18 by 7 to get quotient as 2 and remainder as 4

*   Carry the remainder 4 to the next digit 0 (attached as before) and get 40

*   Multiply the quotient 2 by the flag 7 to get 14 and subtract it from 40 to get 26

*   Divide 26 by 7 to get the quotient as 3 and remainder as 5

*   This process can be carried out upto any desired level of accuracy.

Case 3 : Division with adjustments

Consider the division of 25382 by 77

7								
7	25	3	8	: 2	0		0	0
		4	8	11	7		7	7
		− 21	− 14	− 63	− 42		− 21	− 42
		22	74	49	28		49	28
	3	2	9	: 6	3		6	3

Steps :

- Place a colon one digit from the right (= number of digits of flag)

- Divide 25 by 7 to get quotient as 3 and remainder as 4

- Carry the remainder 4 to the next digit 3 and get **43 (GD)**

- Multiply the quotient 3 by the flag 7 to get 21 and subtract it from 43 to get **22 (ND)**

- Divide 22 by 7 to get the quotient as 3 and remainder as 1

- Carry the remainder 1 to the next digit 8 and get **18 (GD)**

- Multiply the quotient 3 by the flag 7 to get 21 and subtract it from 18 to get –3 **(ND)**

- Now, we have a peculiar situation wherein the 'Net Dividend' has become negative. The process cannot go any further. We will now have to make an adjustment.

- We will go one step backward and instead of writing the quotient as 3 while dividing 22 by 7, we will write down only 2 as the quotient and carry forward a bigger remainder of 8 (i.e. 22 – 7 * 2)

- Carry the remainder 8 to the next digit 8 and get **88 (GD)**

- Multiply the quotient 2 by the flag 7 to get 14 and subtract it from 88 to get **74 (ND)** which is now positive

- Carry out the process in the remaining columns, making an adjustment whenever the net dividend becomes negative.

  The final result is 329 with remainder 49 or in decimal form, it is 329.6363.

Case 4 : Division by a number with a flag of 2 digits

Let us now consider division by a 3-digit divisor where the flag would have two digits.

Consider the division of 2329989 by 514

14							
5	23	2	9	9 :	8	9
		3	3	3	1	3
		- 4	- 21	- 23	-15	-12
		28	18	16	3	27
	4	5	3	3 :	0	5

Steps :

- Place a colon two digits from the right, since we now have two digits in the flag

- Divide 23 by 5 to get quotient as 4 and remainder as 3

- Carry the remainder 3 to the next digit 2 and get **32 (GD)**

- Now, since there are two digits in the flag, we will use two quotient digits from the previous two columns to carry out a cross multiplication with the two digits of the flag. We will then subtract the result from the gross dividend **(GD)** to compute the net dividend **(ND)**.

- Since the quotient 4 has only one digit, we pad it with a zero on the left and multiply the quotient 04 by the flag 14 as follows :

  a) 1   4
  b) 0   4

  $$4 \times 0 + 1 \times 4 = 4$$

- We will subtract 4 from the GD 32 to get **28 (ND)**

- Divide 28 by 5 to get the quotient as 5 and remainder as 3

- Carry the remainder 3 to the next digit 9 and get **39 (GD)**
- Now, the previous two quotient digits are 4 and 5 which are multiplied by the flag 14 as follows :

a) 1   4

b) 4   5

$$1 \times 5 + 4 \times 4 = 21$$

We will subtract 21 from the GD 39 to get **18 (ND)**.

- Repeat the process for the remaining columns.

Case 5 : Division by a number with a flag of 3 digits

Let us now consider division by a 4-digit divisor where the flag would have three digits.

Consider the division of 6851235 by 6524

524							
6	6	8	5	1 :	2	3	5
		0	3	3	2	6	8
		-5	-2	-29	-10	-25	
		3	33	2	12	38	
1	0	5	0 :	1	5		

Steps :

- Place a colon three digits from the right since now there are three digits in the flag
- Divide 6 by 6 to get quotient as 1 and remainder as 0
- Carry the remainder 0 to the next digit 8 and get **08 (GD)**
- Now, since there are three digits in the flag, we will use

three quotient digits from the three immediately previous columns and carry out a cross multiplication with the three digits of the flag. We will then subtract the result from the gross dividend **(GD)** to compute the net dividend **(ND)**.

- Since the quotient 1 has only one digit, we pad it with two zeros on the left and multiply the quotient 001 by the flag 524 as follows:

a) 5   2   4
b) 0   0   1

———————————————

$4 \times 0 + 5 \times 1 + 2 \times 0 = 5$

———————————————

- We will subtract 5 from the GD 8 to get **3 (ND)**
- Divide 3 by 6 to get the quotient as 0 and remainder as 3
- Carry the remainder 3 to the next digit 5 and get **35 (GD)**
- Now, the previous two quotient digits are 0 and 1. We pad them with one zero and multiply by the flag 524 as follows:

a) 5   2   4
b) 0   1   0

———————————————

$4 \times 0 + 5 \times 0 + 2 \times 1 = 2$

———————————————

We will subtract 2 from the GD 35 to get **33** (ND).

- Divide 33 by 6 to get the quotient as 5 and remainder as 3
- Carry the remainder 3 to the next digit 1 and get **31 (GD)**
- Now, the previous three quotient digits are 1, 0 and 5. We multiply by the flag 524 as follows :

a) 5   2   4
b) 1   0   5

———————————————

$4 \times 1 + 5 \times 5 + 2 \times 0 = 29$

We will subtract 29 from the GD 31 to get **2** (ND).

• Repeat the process for the remaining columns.

25. *Examples for Division - Flag of 1 digit*

Sr No	Exercise	Answer
1.	459 / 83	5.5301
2.	6768 / 74	91.4594
3.	2926 / 55	53.2
4.	6481 / 43	150.7209
5.	4974 / 62	80.2258
6.	543 / 76	7.1447
7.	2176 / 43	50.604651
8.	5768 / 54	106.81481
9.	7250 / 49	147.9591
10.	9234 / 52	177.5769
11.	4852 / 68	71.3529
12.	8755 / 83	105.481
13.	1277 / 48	26.6041
14.	3759 / 25	150.36
15.	54225 / 46	1178.8043
16.	49893 / 39	1279.3076
17.	626 / 77	8.1298
18.	92943 / 62	1499.0806
19.	59662 / 880	67.7977
20.	52425 / 56	936.16071

26. Examples for Division - Flag of 2 digits

Sr No	Exercise	Answer
1.	49492 / 572	86.52447
2.	99210 / 495	200.42424
3.	85720 / 734	116.78474
4.	59592 / 422	141.21327
5.	76255 / 942	80.95010
6.	88770 / 542	163.78228
7.	65667 / 349	188.15759
8.	33995 / 662	51.35196
9.	75423 / 550	137.13272
10.	66990 / 339	197.61061
11.	85852 / 458	187.44978
12.	57599 / 399	144.35839
13.	98544 / 624	157.92307
14.	39256 / 199	197.26633
15.	68825 / 525	131.09523
16.	36395 / 636	57.224842
17.	94850 / 540	175.64814
18.	67822 / 426	159.20657
19.	64655 / 348	185.79022
20.	59435 / 459	129.48801

27. Examples for Division - Flag of 3 digits

Sr No	Exercise	Answer
1.	3986549 / 3524	1131.25681
2.	740219 / 2531	292.46108

3.	18352 / 3213	5.711795
4.	129221 / 2142	60.32726
5.	182.321 / 2412	0.0755891
6.	8.53415 / 4.124	2.069386
7.	621356 / 1521	408.51808
8.	628851 / 5221	120.44646
9.	35.9361 / 3.521	10.206219
10.	5.46781 / 3.124	1.750259
11.	72231 / 2331	30.98712
12.	68.52315 / 6542	0.010474
13.	674255 / 3452	195.32300
14.	7.34652 / 2.431	3.022015
15.	10362 / 3123	3.317963
16.	730219 / 2532	288.39612
17.	176.521 / 1524	0.1158274
18.	15396 / 3126	4.9251439
19.	54612 / 2414	22.623032
20.	67543 / 3425	19.720583

CHAPTER 7

Simple Squares

In this chapter, we will look at two simple techniques. In the first one, we will learn how to find the square of a number which ends in 5, eg. 35. In the second one, we will see how to find the product of two numbers which have the same starting digit and which have the sum of the ending digits as 10. eg. 33 and 37.

Case 1 : Numbers ending with 5

Consider the square of 35. The technique is very simple and consists of two steps.

Steps :

- Multiply the first number, viz 3, by its consecutive number i.e. multiply 3 by 4 and write down the result as 12.

- Next, multiply the second digit, i.e. 5 by itself to obtain 25 and attach it at the end of the result obtained in the previous step.

The result is **1225**.

Example - 1

Compute 65^2

Step 1 : $6 \times 7 = 42$

Step 2 : $5 \times 5 = 25$

And the final answer is **4225**.

Example - 2

Let's compute 125^2

Step 1 : $12 \times 13 = 156$

Step 2 : $5 \times 5 = 25$

And the final answer is **15625**.

The product of 12×13 can be obtained by the Nikhilam method as explained before.

$$
\begin{array}{r}
12 \quad +2 \\
13 \quad +3 \\
\hline
15 / 6 \quad = 156
\end{array}
$$

28. Examples for Squares of numbers ending in 5

Sr No	Exercise	Answer
1.	45^2	2025
2.	165^2	27225
3.	85^2	7225
4.	95^2	9025
5.	6.5^2	42.25
6.	0.75^2	0.5625
7.	55×5500	302500
8.	85×0.0085	0.7225
9.	0.035×3.5	0.1225
10.	9.85^2	97.0225

11.	1.75^2	3.0625
12.	22.35×2.235	49.95225
13.	9925^2	9805625
14.	18.5^2	342.25
15.	19.5^2	380.25
16.	625^2	390625
17.	14.5^2	210.25
18.	11.35^2	128.8225
19.	100.5^2	10100.25
20.	0.0115^2	0.00013225

Case 2 : Two numbers starting with same digit and ending digits adding up to 10

Consider the example 23 × 27.

Now, 23 × 27 = 621

Let's see how this product can be obtained by using the same technique as for squaring numbers ending with 5.

Steps :

• Since both the numbers start with the same digit, multiply it (i.e. 2) by its consecutive number (i.e. 3) to get 6.

• Now multiply the last digits of both the numbers viz 3 × 7 to get 21.

The final result is obtained by the concatenation of the two numbers i.e. '6' + '21' = **621**.

Example - 1

Compute 84 × 86

The result will be obtained as

Step 1 : $8 \times 9 = 72$ (first digit '8' multiplied by its next digit '9')

Step 2 : $4 \times 6 = 24$ (product of last digit of each number) giving the final result as **7224.**

Example - 2

Compute 92×0.98

The result can be obtained as

Step 1 : $9 \times 10 = 90$

Step 2 : $2 \times 8 \quad = 16$

which gives the intermediate answer as 9016.

Now, place the decimal point after two digits from the right to get the final answer as **90.16.**

Important

The same technique can also be used for bigger numbers like 318×312 or 1236×1234, where the last digits add up to 10 and the remaining digits are the same.

29. Examples for Product of numbers having ending digits adding to 10

Sr No	Exercise	Answer
1.	26×24	624
2.	33×37	1221
3.	81×89	7209
4.	93×97	9021
5.	48×42	2016

6.	5.6 × 5.4	30.24
7.	0.69 × 6.1	4.209
8.	9.3 × 0.0097	0.09021
9.	0.71 × 0.79	0.5609
10.	0.82 × 880	721.60
11.	1.1 × 19	20.90
12.	0.12 × 0.0018	0.000216
13.	430 × 0.47	202.10
14.	117 × 11.3	1322.10
15.	72 × 7800	561600
16.	8.7 × 83	722.10
17.	71 × 79	5609
18.	340 × 36	12240
19.	9.4 × 0.96	9.024
20.	88 × 0.82	72.16

CHAPTER 8

Square of any number

There are many problems in which we have to compute the square of a number, i.e. multiply a number by itself. Eg. square of 382, 2.3672 etc.

Vedic maths provides a powerful method to compute the square of any number of any length with ease and get the answer in one line. It uses the concept of 'dwanda' or duplex which is explained in this chapter.

We will start with some basic definitions, which are the building blocks for these techniques.

I. Definition - Dwandwa or Duplex

We will define a term called 'dwandwa' or duplex denoted by 'D'.

a) Duplex of a single digit is defined as

$$D(a) = a^2$$

Thus $D(3) = 3^2 = 9,$

$$D(6) = 6^2 = 36 \text{ etc.}$$

b) Duplex of two digits is defined as

$$D(ab) = 2 * a * b$$

Thus, $D(37) = 2 * 3 * 7 = 42$

$$D(41) = 2 * 4 * 1 = 8$$

$$D(20) = 0$$

c) *Duplex of 3 digits is defined as*

$D(abc) = 2 * a * c + b^2$

This can be derived by using $D(a)$ and $D(ab)$ defined above.
We pick up the digits at the two extreme ends i.e. 'a' and 'c' and compute its duplex as $2 * a * c$.

We then move inwards and pick up the remaining digit 'b'.
We add the duplex of b, i.e. b^2 to the result to get the duplex of the 3 digits 'a', 'b' and 'c'.

Thus $D(346) = 2 * 3 * 6 + 4^2$
$$= 36 + 16 = 52$$
$$D(130) = 2 * 1 * 0 + 3^2 = 9$$
$$D(107) = 2 * 1 * 7 + 0^2 = 14.$$

d) *Duplex of 4 digits is defined as*

$D(abcd) = 2 * a * d + 2 * b * c = 2 (a * d + b * c)$

Once again, we start with the two digits 'a' and 'd', compute their duplex as $2 * a * d$, then move inwards, we get another pair 'b' and 'c' and compute their duplex as $2 * b * c$.

Thus the final duplex is $2 * a * d + 2 * b * c$

$D(2315) = (2 * 2 * 5) + (2 * 3 * 1)$
$$= 20 + 6 = 26$$
$D(3102) = (2 * 3 * 2) + (2 * 1 * 0)$
$$= 12 + 0 = 12$$
$D(5100) = (2 * 5 * 0) + (2 * 1 * 0)$
$$= 0$$

e) *Duplex of five digits is defined as*

$D(abcde) = 2ae + 2bd + c^2$
$D(21354) = (2 \times 2 \times 4) + (2 \times 1 \times 5) + 3^2$
$$= 16 + 10 + 9 = 35$$
$D(31015) = (2 \times 3 \times 5) + (2 \times 1 \times 1) + 0^2$
$$= 30 + 2 + 0 = 32$$

Summary of computation of Duplex

Duplex of	Required Definition
D(a)	a^2
D(ab)	$2ab$
D(abc)	$2ac + b^2$
D(abcd)	$2(ad + bc)$
D(abcde)	$2(ae + bd) + c^2$

30. Compute the duplex of the following numbers

Sr No	Type	Exercise	Answer
1.	2-Digit Numbers	34, 47, 23, 10, 70	24, 56, 12, 0, 0
2.	3-Digit Numbers	132, 102, 120, 151, 100	13, 4, 4, 27, 0
3.	4-Digit Numbers	1324, 2114, 2405, 2001, 2000	20, 18, 20, 4, 0
4.	5-Digit Numbers	32513, 30725, 32025, 30035, 30005, 30125, 40100, 41111	47, 79, 38, 30, 30, 31, 1, 11

II. Square of any Number

Once, the concept of 'duplex' is clear, we can compute the square of any number very easily. Let us start with the square of a 2-digit number and then extend it to bigger numbers.

The square of a 2-digit number ab is defined as
$(ab)^2 = D(a) / D(ab) / D(b)$

Consider the square of 34.

Then, $34^2 = D(3) / D(34) / D(4)$

where $D(3) = 9$, $D(34) = 24$ and $D(4) = 16$

So, $34^2 = 9/24/16$.

Now, we start a backward pass from the right, retain one digit at each location and carry the surplus to the left. The working would look as follows :

$34^2 = 9 / 24 / 16$
$= 9 / 25 / 6$ .. We retain 6 and carry 1 to the left getting 25.
$= 11 / 5 / 6$ .. We retain 5 and carry 2 to the left getting 11.

So the final answer is 1156.

Similarly, $82^2 = 64/32/4$
$= 6724$

The technique can now be extended to a 3-digit number as follows :

$(abc)^2 = D(a) / D(ab) / D(abc) / D(bc) / D(c)$
Thus, $213^2 = 4 / 4 / 13 / 6 / 9$
$= 45369$

The square of any number can now be obtained very easily by just extending the set of duplexes.

31. Examples for Squares of 2-digit numbers

Sr No	Number	Square Value
1.	32	1024
2.	53	2809
3.	62	3844
4.	88	7744
5.	92	8464
6.	96	9216
7.	2.8	7.84

8.	9.4	88.36
9.	49	2401
10.	3.4	11.56
11.	0.53	0.2809
12.	0.066	0.004356
13.	58	3364
14.	77	5929
15.	0.0054	0.00002916
16.	59	3481
17.	0.67	0.4489
18.	87	7569
19.	0.43	0.1849
20.	0.64	0.004096

32. Squares of 3-Digit Numbers

Sr No	Number	Computation of Duplex	Square Value
1.	212	4/4/9/4/4	44944
2.	315	9/6/31/10/25	99225
3.	239	4/12/45/54/81	57121
4.	227	4/8/32/28/49	51529
5.	515	25/10/51/10/25	265225
6.	733	49/42/51/18/9	537289
7.	779	49/98/175/126/81	606841
8.	649	36/48/124/72/81	421201
9.	23.1	4/12/13/6/1	533.61
10.	32.6	9/12/40/24/36	1062.76
11.	44.7	16/32/72/56/49	1998.09

12.	0.0731	49/42/23/6/1	0.00534361
13.	0.000232	4/12/17/12/4	.000000053824
14.	0.00431	16/24/17/6/1	.0000185761
15.	4.78	16/56/113/112/64	22.8484
16.	756	49/70/109/60/36	571536
17.	434	16/24/41/24/16	188356
18.	0.0672	36/84/63/28/4	0.0045158
19.	4.48	16/32/80/64/64	20.0704
20.	616	36/12/73/12/36	379456

33.Examples for Squares of 4-Digit Numbers

Sr No	Number	Computation of Duplex	Square Value
1.	2103	4/4/1/12/6/0/9	4422609
2.	1333	1/6/15/24/27/18/9	1776889
3.	1231	1/4/10/14/13/6/1	1515361
4.	2419	4/16/20/44/73/18/81	5851561
5.	5706	25/70/49/60/84/0/36	32558436
6.	4213	16/16/12/28/13/6/9	17749369
7.	7939	49/126/123/180/171/54/81	63027721
8.	6677	36/72/120/168/133/98/49	44582329
9.	1.352	1/6/19/34/37/20/4	1.827904
10.	2.201	4/8/4/4/4/0/1	4.844401
11.	32.03	9/12/4/18/12/0/9	1025.9209
12.	0.2332	4/12/21/26/21/12/4	0.05438224
13.	3.014	9/0/6/24/1/8/16	9.084196
14.	0.03251	9/12/34/26/29/10/1	0.0010569001
15.	4.023	16/0/16/24/4/12/9	16.184529

16.	40.82	16/0/64/16/64/32/4	1666.2724
17.	1212	1/4/6/8/9/4/4	1468944
18.	41.13	16/8/ 9/26/7/6/ 9	1691.6769
19.	0.2321	4/12/17/16/10/4/1	0.05387041
20.	1246	1/4/12/28/40/48/36	1552516

34. Examples for Squares of 5-Digit Numbers

Sr No	Number	Computation of Duplex	Square Value
1.	32141	9/12/10/28/23/12/18/8/1	1033043881
2.	31243	9/6/13/28/30/22/28/24/9	976125049
3.	10543	1/0/10/8/31/40/46/24/9	111154849
4.	52143	25/20/14/44/47/20/22/24/9	2718892449
5.	61524	36/12/61/34/77/28/44/16/16	3785202576
6.	215.26	4/4/21/18/53/32/64/24/36	46336.8676
7.	239.21	4/12/45/62/97/42/22/4/1	57221.4241
8.	30981	9/0/54/48/87/144/82/16/1	959822361
9.	36.098	9/36/36/54/156/96/81/144/64	1303.065604
10.	63254	36/36/33/72/82/44/41/40/16	4001068516
11.	209.87	4/0/36/32/109/144/190/112/49	44045.4169
12.	12179	1/4/6/18/47/50/67/126/81	148328041
13.	153.67	1/10/31/42/83/106/78/84/49	23614.4689
14.	53206	25/30/29/12/64/36/24/0/36	2830878436
15.	33032	9/18/9/18/30/12/9/12/4	1091113024
16.	10612	1/0/12/2/40/12/5/4/4	112614544
17.	2.3121	4/12/13/14/17/10/6/4/1	5.34580641
18.	21.433	4/4/17/20/34/30/33/18/9	459.373489
19.	15369	1/10/31/42/87/126/90/108/81	236206161
20.	30797	9/0/42/54/91/126/179/126/49	948455209

CHAPTER 9

Square Root of a number

In the previous chapter, we have seen the concept of 'duplex' and how it helps us to compute the square of any number very rapidly.

The computation of the square root also involves the use of the duplex, only this time, we would be subtracting it at each step to compute the square root.

We will now study the technique to find the square root of any number which may or may not be a perfect square. The first three examples are of numbers, which are perfect squares and the next two examples are of imperfect squares including a decimal number. The general steps are explained with specific examples below :

Steps :

• Count the number of digits, n, in the given number.

• If the number of digits is even, start with the leftmost two digits, else start with the leftmost single digit. Let us call the number so obtained as 'SN' (Starting Number).

• Consider the highest possible perfect square, say 'HS', less than or equal to 'SN' and compute its square root 'S'.

• Write down the square root 'S', below the leftmost digits. This becomes the first digit of the required square root.

- Multiply the square root 'S' by 2 and write it on the left side. This would be the divisor, say 'd'.

- For all the balance digits in the original number, the method would consist of division by the divisor 'd' after subtraction of the duplex of the previous digits (except the first one).

- The difference between 'SN' and 'HS' i.e. the remainder 'R' is carried to the next digit and written to its lower left, forming a new number. This next number is called the gross dividend (GD).

- The first GD is divided by the divisor 'd' to get a quotient Q and a remainder 'R'. The quotient is written down as the second digit of the square root and the remainder is again carried to the next digit to form the next GD.

- From this stage onwards, the duplex of all digits starting from the second digit of the square root, to the column previous to the column of computation is subtracted from each GD to get the Net Dividend, ND.

- At each column, the ND is then divided by the divisor 'd' and the quotient is written down as the next digit of the square root and the remainder is carried to the next digit to form the next GD.

- The process is carried out till the ND becomes zero and no more digits are left in the number for computation or when we have reached a desired level of accuracy.

- If the ND does not become zero, we can attach zeroes to the end of the given number and carry out the process to get the square root to any level of accuracy, i.e. any number of decimal places.

- Once the square root is obtained or when we decide to stop, we place the decimal point after the required digits from the left. If n is even, the decimal point is put (n / 2) digits from the left. If n is odd, the decimal point is placed (n + 1) / 2 digits from the left.

The process will become very clear on studying the examples given below.

I. Perfect square root

Consider the example of computing the square root of 2116

The working is as follows which is explained stepwise.

```
        21    1      6
              5      3
    8              - 36
                       0
        4     6  .   0
```

Steps :

- Count the number of digits, n, which is 4.

- Since n is even, we start with the two leftmost digits where 'SN' is 21.

- Here the highest perfect square ('HS') less than 21 is 16, its square root ('S') is 4 and the difference 'R' is 5.

- We write the square root 4 below the leftmost digits and write 8 (= 2 × 4) on the left side as shown.

- 4 is the first digit of the required square root.

- For the balance digits, the method would consist of division by the divisor 8.

- The difference i.e. 5 is carried to the next digit 1 and written to its lower left as shown. The value of the GD is now 51.

- The very first GD i.e. 51 is divided by the divisor 8 to get a quotient 6 and a remainder 3. The quotient is written down as part of the final result and the remainder is carried to the next digit 6 to form the next GD 36.

- We have now reached the third column. We will subtract the duplex of the digit in the second column from the GD to get the Net Dividend, ND.

- The digit in the second column of the result is 6 and its duplex is 36.

- 36 is subtracted from the GD which is also 36 to get the ND as 0. This signals the end of the process.

- Hence, the square root is obtained. Now, we will place the decimal point.

- Since, n is 4 and is even, the decimal point is placed (n / 2) digits from the left. So, here, the decimal point will be placed 2 digits from the left and the final answer is 46.

II. Perfect square root

Consider the example of computing the square root of 59049

	5	9	0	4	9
		1	3	2	0
4			- 16	-24	-9
			14	*0*	*0*
	2	4	3	. 0	0

Steps :

• Count the number of digits, n, which is 5.

• Since n is odd, we start with the left-most digit viz 5

• Here the highest perfect square less than 5 is 4, its square root is 2 and the remainder 'R' is 1 (5 − 4).

• We write the square root 2 below the left-most digit and write 4 (= 2 × 2) on the left side as shown.

• The remainder 1 is carried to the next digit 9 giving the value of the GD as 19.

• 19 is divided by 4 to get a quotient 4 and a remainder 3. The quotient is written down and the remainder is carried to the next digit 0 to give the next GD as 30.

• The duplex of 4 is 16 which is subtracted from 30 (GD) to get 14 (ND).

• 14 is divided by 4 to get a quotient 3 and a remainder 2. The quotient is written down and the remainder is carried to the next digit 4 to give the next GD **24**.

• Now, the duplex of the two previous digits 4 and 3, i.e. 2 * 4 * 3 = 24 is subtracted from the GD 24 to get the ND as 0. Since, we have not yet reached the last digit in the given number, we cannot stop here.

• On dividing 0 by 4 we get the quotient as 0 and the remainder as 0, which is carried to the next digit giving the GD as 09.

• The three immediately previous digits are 4, 3 and 0. Their duplex is 9 (2 * 4 * 0 + 3²). When the duplex is subtracted, the ND becomes equal to zero and no more digits are left. This signals the end of the process.

• Since, n is 5, the decimal point is placed (n + 1) / 2 digits i.e. 3 digits from the left and the final answer is 243.

III. Imperfect square root as a decimal number

Consider the example of computing the square root of 59064

Since this number is slightly larger than the number taken in the previous example, it is not a perfect square. We will now see how to get its square root to any desired level of accuracy. At any point, if the ND becomes negative, we go back one step and decrease the quotient by one to increase the remainder that is carried forward. This process of adjustment is explained in this example.

	5	9	0	6	4	0	0	0
		1	3	2	2	3	6	10
4			-16	-24	-9	-24	-18	-64
			14	2	15	6	42	36
	2	4	3 .	0	0	0	8	6

Steps :

• The first few steps are the same till we reach the fourth column where the GD is 26 and the ND is 2.

• 2 is divided by 4 to get a quotient 0 and a remainder 2. The quotient is written down and the remainder is carried to the next digit 4 to give the next GD 24.

• The duplex of 4, 3 and 0 is 9 which is subtracted from 24 (GD) to get 15 (ND).

• 15 is divided by 4 to get a quotient 3 and a remainder 3. We attach a zero to continue with the process.

- The quotient is written down and the remainder is carried to the next digit 0 to form the next GD **30**.

- Now, the duplex of the previous digits 4, 3, 0 and 3, i.e. 2 * 4 * 3 = 24 is subtracted from the GD 30 to get the ND as 6.

- On dividing 6 by 4 we get the quotient as 1 and the remainder as 2, which is carried to the next digit giving the GD as 20.

- The duplex of the immediately previous digits 4, 3, 0, 3 and 1 is 26 (2 * 4 * 1 + 2 * 3 * 3) which gives a negative ND. Hence, we decrease the previous quotient by 1 and carry forward the remainder 6 giving the GD as 60.

- Now the duplex of the previous digits 4, 3, 0, 3 and 0 is 18 (2 * 3 * 3). When this duplex is subtracted from the GD, we get the ND as 42.

- This process can be carried out to any desired level of accuracy.

- Since n is 5, the decimal point is put [(5 + 1) / 2] = 3 digits from the left, giving the final answer as 243.03086.

IV. Square root with adjustments
Consider the square root of a decimal number 59064.78

This is a decimal number which is not a perfect square.

4	5	9	0	6	4	.	7	8	0	0
		1	3	2	2		3	5	8	12
			-16	-24	-9		-24	-34	-44	-81
			14	2	15		13	24	36	39
	2	4	3	.	0		3	2	4	6	7

Steps :

- Write the number without the decimal point and carry out the same steps as explained before.

- The starting number (SN) is obtained from the integer port only of the decimal number.

- The first few steps are the same till we reach the fifth column where the GD is 24 and the ND is 15.

- As before, 15 is divided by 4 to get a quotient 3 and a remainder 3. We carry the remainder to the next digit 7 to get 37 (GD).

- Now, the duplex of the previous digits 4, 3, 0 and 3 is 24 (2 * 4 * 3) which is subtracted from the GD 37 to get the ND as 13.

- On dividing 13 by 4, we get the quotient as 2 and the remainder as 5, which is carried to the next digit giving the GD as 58. Note that the quotient is taken as 2 and not 3 to get a positive ND in the next column.

- The duplex of the immediately previous digits 4, 3, 0, 3 and 2 is 34 (2 * 4 * 2 + 2 * 3 * 3) which is subtracted from 58 giving the ND as 24.

- 24 is divided by 4 to get a quotient of 4 and a remainder of 8 which is carried forward. A zero is attached at this stage to enable the process to continue to higher levels of accuracy.

- This process is carried further as shown

- Since, n is 5, the decimal point is put (5 + 1) / 2 = 3 digits from the left, giving the final answer as 243.032467

V. Perfect square root of a decimal number

Now, consider the square root of a decimal number 547.56 which is a perfect square.

```
        5      4      7      5      6
               1      2      2      1
    4        - 9    -24    -16
                     18      1      0
        2      3  .  4      0      0
```

Steps:

*   The steps are similar to those explained before and the reader should have no difficulty in computing the same.

*   Since the original number has 3 digits, the decimal point is placed after 2 digits from the left.

Important

If the starting number SN is small, say less than 4, then the next group of 2 digits can be included in the SN. Eg in computing the square root of 13251, we can take SN as 132 instead of 1.

35. Examples for Square Roots of Miscellaneous numbers

Sr No	Number	Square Root
1.	14641	121
2.	800.89	28.3
3.	17.7241	4.21
4.	506944	712
5.	22881	151.26

6.	1373584	1172
7.	3981.61	63.1
8.	919681	959
9.	264196	514
10.	973.44	31.2
11.	125.8884	11.22
12.	674041	821
13.	5.349969	2.313
14.	135424	368
15.	29062881	5391
16.	761.76	27.6
17.	15376	124
18.	133956	366
19.	534361	731
20.	203401	451

CHAPTER **10**

CHAPTER **10**

Cubes and Cube Roots

a) Computing cubes of 2-digit numbers

Let us consider a two digit number say 'ab'. Further, let us consider the expansion of the expression $(a + b)^3$, which can be used to find the cube of the given number 'ab'.

We know that $(a + b)^3 = a^3 + 3a^2b + 3ab^2 + b^3$

We notice that the 1st term is a^3,
the 2nd term $a^2b = a^3 \times (b/a)$,
the 3rd term $ab^2 = a^2b \times (b/a)$ and
the 4th term $b^3 = ab^2 \times (b/a)$

Thus, each of the 2nd, 3rd and 4th term can be obtained from its previous term by multiplying it by the common ratio (b/a).

We can also consider the 2nd term $3a^2b = a^2b + 2a^2b$
and the 3rd term as $3ab^2 = ab^2 + 2ab^2$

i.e. we can split it as the sum of two terms.

Hence, if we compute a^3 and the ratio (b/a), we can derive all the remaining terms very easily. The two middle terms have to be then doubled and added as shown to get $(ab)^3$.

$$(a + b)^3 = a^3 + a^2b + ab^2 + b^3$$
$$+ 2a^2b + 2ab^2$$

$$a^3 + 3a^2b + 3ab^2 + b^3$$

Hence, in order to get the cube of any two digit number, we start by computing the cube of the first digit and the ratio of the 2nd digit by the 1st digit. The ratio can be more than, equal to or less than one. The method remains the same in all cases.

Case 1 : Ratio is more than 1

Let us consider a simple example i.e. 12^3

Here the first digit is 1 and its cube is also 1. The ratio of the 2nd to the 1st digit is 2 (= 2/1).

The remaining 3 terms can be obtained by multiplying each of the previous term by 2 as shown :

$12^3 = 1\ 2\ 4\ 8$ (cube of last digit)

We also notice that the 4th term is the cube of the 2nd digit, i.e. 8 is the cube of 2 which is the second digit in the given number 12.

We have to now take twice the value of each of the two middle terms and write them down. On addition, we get the cube of the given number 12.

12^3 =   1   2   4   8   (cube of last digit)
                   4   8   (double the 2 middle digits)

            1   6 $_1$2   8

     =   1   7   2   8

Similarly, let us look at 25^3

25^3	=	8	20	50	125	(Common ratio = 5/2)
			40	100		

	8	60	150	125	
	8	60	150	5	
				₁₂	

$$8 \quad 60 \quad {}_{16}2 \quad 5 \qquad \text{12 is added to 150 to get 162}$$

$$8 \quad {}_{7}6 \quad 2 \quad 5 \qquad \text{16 is added to 60 to get 76}$$

$$15 \quad 6 \quad 2 \quad 5 \qquad \text{7 is added to 8 to get 15}$$

$$= \quad 15625$$

Case 2 : Ratio is equal to 1

11^3	=	1	1	1	1	(Common ratio = 1)
			2	2		

	1	3	3	1	

$$= \quad 1331$$

Case 3 : Ratio is less than 1

52^3	=	125	50	20	8	(Common ratio = 2/5)
			100	40		

$$140 \quad {}_{15}6 \quad {}_{6}0 \quad 8$$

$$= \quad 140608$$

36. Examples for Cubes of numbers

Sr No	Number	First Line	Answer
1.	13^3	1, 3, 9, 27	2197
2.	15^3	1, 5, 25, 125	3375
3.	17^3	1, 7, 49, 343	4913
4.	0.019^3	1, 9, 81, 729	0.000006859
5.	2.2^3	8, 8, 8, 8	10.648
6.	32^3	27, 18, 12, 8	32768
7.	2.3^3	8, 12, 18, 27	12.167
8.	27^3	8, 28, 98, 343	19683
9.	26^3	8, 24, 72, 216	17576
10.	0.45^3	64, 80, 100, 125	0.091125
11.	63^3	216, 108, 54, 27	250047
12.	28^3	8, 32, 128, 512	21952
13.	470^3	64, 112, 196, 343	103823000
14.	56^3	125, 150, 180, 216	175616
15.	51^3	125, 25, 5, 1	132651
16.	18^3	1, 8, 64, 512	5832
17.	4.6^3	64, 96, 144, 216	97.336
18.	105^3	1000, 500, 250, 125	1157625
19.	111^3	1331, 121, 11, 1	1367631
20.	0.88^3	512, 512, 512, 512	0.681472

b) Cube roots of 2 digit numbers

Let's consider the cubes of numbers from 1 to 9.

$1^3 = 1$
$2^3 = 8$
$3^3 = 27$

$4^3 = 64$
$5^3 = 125$
$6^3 = 216$
$7^3 = 343$
$8^3 = 512$
$9^3 = 729$

If we observe closely, we find that the last digit in every cube is unique and does not repeat for any number (1 to 9). This property can be used to find the cube roots of any two digit numbers very easily.

Let's find the cube root of 438976.

We start from the right and group it into sets of 3 digits each. So, we have two groups viz
438   976

Steps :

*   We look at the first group from the right, i.e. 976 and compare its last digit, i.e. 6 with the cubes of digit from 1 to 9 where the cube also ends with 6. The number 216 ends with 6.

*   We take its cube root which is 6. The last digit of the required cube root is 6.

*   Now, we consider the remaining 3-digit group i.e. 438. We locate the highest cube less than or equal to 438. The required cube is 343 since the cube higher than it is 512 which is greater than 438.

*   We take its cube root which is 7. The first digit of the required cube root is 7.

*   The required cube root is 76.

This technique provides an easy and powerful method to

compute 2-digit cube roots. It can be used only when the given number is a perfect cube.

37. Examples for Cube roots of 2 digit numbers

Sr No	Number	Cube Root
1.	175616	56
2.	238328	62
3.	50653	37
4.	250.047	6.3
5.	157464	54
6.	474552	78
7.	148877	53
8.	0.287496	0.66
9.	571787	83
10.	830584	94
11.	941192	98
12.	103823	47
13.	54872	38
14.	32768	32
15.	438976	76
16.	704969	89
17.	185193	57
18.	0.636056	0.86
19.	110592	48
20.	59319	39

c) Computing fourth power of 2 digit numbers

A similar approach can be used to obtain the fourth power of a two digit number, i.e. $(ab)^4$

$$(a + b)^4 = a^4 + 4a^3b + 6a^2b^2 + 4ab^3 + b^4$$

$$= a^4 + a^3b + a^2b^2 + ab^3 + b^4 \text{ (Ratio = b/a)}$$

$$3a^3b + 5a^2b^2 + 3ab^3$$

$$a^4 + 4a^3b + 6a^2b^2 + 4ab^3 + b^4$$

Subsequent to the first term, each of the remaining terms is obtained by multiplying the previous term by b/a. Let us consider examples from each of the cases viz, when the ratio is more than, equal to and less than 1.

Case 1 : Ratio is more than 1

Let us consider a simple example, i.e. 12^4

Here the first digit is 1 and its fourth power is also 1. The ratio of the 2nd to the 1st digit is 2.

The remaining 4 terms can be obtained by multiplying each of the previous term by 2 as shown :

$$12^4 = \quad 1 \quad 2 \quad 4 \quad 8 \quad 16$$
$$6 \quad 20 \quad 24$$

1	8	24	32	16	
1	8	24	32	$_16$	.. Add 1 to left giving 33
1	8	24	$_33$	6	.. Add 3 to left giving 27
1	8	$_27$	3	6	.. Add 2 to left giving 20
2	0	7	3	6	

$$= 20736$$

Case 2 : Ratio is equal to 1

Let us consider a simple example, i.e. 11^4

Here the first digit is 1 and its fourth power is also 1. The ratio of the 2nd to the 1st digit is 1.

The remaining 4 terms can be obtained by multiplying each of the previous term by 1 as shown :

$$11^4 = \begin{array}{ccccc} 1 & 1 & 1 & 1 & 1 \\ & 3 & 5 & 3 & \\ \hline 1 & 4 & 6 & 4 & 1 \end{array}$$

$$= 14641$$

Case 3 : Ratio is less than 1

Let us consider a simple example, i.e. 21^4

Here the first digit is 2 and its fourth power is 16. The ratio of the 2nd to the 1st digit is 1/2.

The remaining 4 terms can be obtained by multiplying each of the previous term by 1/2 as shown :

$$21^4 = \begin{array}{ccccc} 16 & 8 & 4 & 2 & 1 \\ & 24 & 20 & 6 & \\ \hline 19 & 4 & 4 & 8 & 1 \end{array}$$

$$= 194481$$

38. Examples for fourth power of numbers

Sr No	Number	First Line	Answer
1.	13^4	1, 3, 9, 27, 81	28561
2.	15^4	1, 5, 25, 125, 625	50625
3.	17^4	1, 7, 49, 343, 2401	83521
4.	0.19^4	1, 9, 81, 729, 6561	0.00130321
5.	2.2^4	16, 16, 16, 16, 16	23.4256
6.	32^4	81, 54, 36, 24, 16	1048576

7.	2.3^4	16, 24, 36, 54, 81	27.9841
8.	27^4	16, 56, 196, 686, 2401	531441
9.	26^4	16, 48, 144, 432, 1296	456976
10.	0.45^4	256, 320, 400, 500, 625	0.4100625
11.	63^4	1296, 648, 324, 162, 81	15752961
12.	28^4	16, 64, 256, 1024, 4096	614656
13.	470^4	256, 448, 784, 1372, 2401	48796810000
14.	56^4	625, 750, 900, 1080, 1296	9834496
15.	51^4	625, 125, 25, 5, 1	6765201
16.	0.12^4	1, 2, 4, 8, 16	0.00020736
17.	5.3^4	625, 375, 225, 135, 81	789.0481
18.	330^4	81, 81, 81, 81, 81	11859210000
19.	29^4	16, 72, 324, 1458, 6561	707281
20.	4.2^4	256, 128, 64, 32, 16	311.1696

CHAPTER 11

Auxiliary Fractions

We have seen elegant techniques provided by vedic maths for division of any number by any divisor. The methods make the process easy and fast and coupled with the techniques of mishrank (Chapter 12), the increase in speed is amazing.

We will now see additional techniques which can speed up the division of fractional numbers where :

- the dividend is less than the divisor and
- the divisor is small, e.g. having 2 or 3 digits and
- the last digit of the divisor ends with 1, 6, 7, 8 and 9

    Eg. of such divisors are 19, 29, 27, 59, 41, 67, 121 etc.

Other divisors ending in 2, 3, 4, 5 and 7 can be converted to such divisors by multiplying by a suitable number.

We will see examples of divisors ending in each of the digits from 1 to 9 to get a good insight into the entire process.

I) Divisors ending with '9'

Let us start with the case where the divisor ends with 9. Eg. 19, 29, 39, 89 etc.

General Steps

- Add 1 to the denominator and use that as the divisor

- Remove the zero from the divisor and place a decimal point in the numerator at the appropriate position

- Carry out a step-by-step division by using the new dividend,

- Every division should return a quotient and a remainder and should not be a decimal division.

- Every quotient digit will be used without any change to compute the next number to be taken as the dividend. In all other cases, where the divisor does not end with 9 the quotient digit would be altered by a pre-defined method which will be explained in each case.

Example 1 :

Let us now see the details for 5 / 29

Steps

- Add 1 to the divisor to get 30

- The modified division is now 5 / 30

- Remove the zero from the denominator by placing a decimal point in the numerator giving the modified division as 0.5 / 3

- We will not write the final answer as 0.16666 but

- We will carry out a step-by-step division explained below.

Steps for step-by-step division

- Divide 5 by 3, to get a quotient (q) = 1 and

  remainder (r) = 2

- Write down the quotient in the answer line and the remainder just below it to its left as shown

```
0.5 / 3 = 0.  1
              2
```

Now, the next number for division is 21 which on division by 3 gives q = 7 and r = 0. The result now looks as

```
0.5 / 3 = 0.  1    7
              2    0
```

Repeat the division at each digit to get the final answer to any desired level of accuracy!! The next few digits are shown below.

On dividing 7 by 3, we get q = 2 and r = 1. The result now looks as

```
0.5 / 3 = 0.  1    7    2
              2    0    1
```

On dividing 12 by 3, we get q = 4 and r = 0. The result now looks as

```
0.5 / 3 = 0.  1    7    2    4
              2    0    1    0
```

The final result upto 8 digits of accuracy is

```
0.5 / 3 = 0.  1    7    2    4    1    3    7    9
              2    0    1    0    1    2    2    0
```

and the result of 5/29 = 0.17241379.

Example 2 :

Similarly, 88/89 will be computed as

```
88 / 89 = 88 / 90 = 8.8 / 9

8.8 / 9 = 0.  9    8    8    7    6    4    0    4    4    9
              7    7    6    5    3    0    4    4    8    3
```

Example 3 :

Similarly, 41/119 will be computed as

$$41 / 119 = 41 / 120 = 4.1 / 12$$

$$41 / 12 = 0. \quad 3 \quad 4 \quad 4 \quad 5 \quad 3 \quad 7 \quad 8 \quad 1$$
$$ 5 \quad 5 \quad 6 \quad 4 \quad 9 \quad 9 \quad 1 \quad 6$$

So, 41 / 119 = 0.34453781

Notes :

• For all numbers ending with 3, we can convert them to numbers ending with 9 by multiplying by 3.

  Eg. if the divisor is 13, we can convert it to 39 by multiplying by 3.

  Hence, 3/13 = 9/39
  5/23 = 15/69.

• Similary, for numbers ending with 7, we can multiply by 7 to get a number ending with 9.

  Eg. if the divisor is 17, we can convert it to 119 by multiplying by 7.

  Hence, 5/7   = 35/49
  4/17 = 28/119
  5/27 = 35/189.

39. *Examples for Auxiliary Fractions - Divisor ending with 9*

Sr No	Number	Converted to	Answer
1.	1/19 = 1/20	0.1/2	0.052631578947
2.	3/19 = 3/20	0.3/2	0.157894736842

3.	1/29 = 1/30	0.1/3	0.034482758621
4.	4/29 = 4/30	0.4/3	0.137931034483
5.	6/29 = 6/30	0.6/3	0.206896551724
6.	71/89 = 71/90	7.1/9	0.797752808989
7.	1/43 = 3/129 = 3/130	0.3/13	0.023255813953
8.	1/7 = 7/49 = 7/50	0.7/5	0.142857142857
9.	2/39 = 2/40	0.2/4	0.051282051282
10.	3/13 = 9/39 = 9/40	0.9/4	0.230769230769
11.	1/23 = 3/69 = 3/70	0.3/7	0.043478260870
12.	2/53 = 6/159 = 6/160	0.6/16	0.037735849057
13.	1/17 = 7/119 = 7/120	0.7/12	0.058823529412
14.	5/69 = 5/70	0.5/7	0.072463768116
15.	3/17 = 21/119 = 21/120	2.1/12	0.176470588235
16.	0.7/29 = 0.1 * (7/30)	0.1 * (0.7/3)	0.024137931034
17.	2/19 = 2/20	0.2/2	0.105263157895
18.	31/149 = 31/150	3.1/15	0.208053691275
19.	6/59 = 6/60	0.6/6	0.101694915254
20.	5/6.9 = 50/69	10 * (5/69)	0.724637681159

II. Divisors ending with '1'

Now, let us consider the fractions where the divisors are numbers ending with 1.

Eg. 21, 31, 51, 71 etc.

General Steps

- Add 1 to the numerator
- Subtract 1 from the denominator and use that as the divisor
- Remove the zero from the divisor and place a decimal point in the numerator at the appropriate position

- Carry out a step-by-step division by using the new divisor,

- Every division should return a quotient and a remainder and should not be a decimal division.

- Subtract the quotient from 9 at each step to form the next dividend.

Example 1 :

Let us now see the details for 4 / 21

Steps

- Subtract 1 to the numerator to get 3

- Subtract 1 from the divisor to get 20

- The modified division now is 3 / 20

- Remove the zero from the denominator by placing a decimal point in the numerator

- The modified division is 0.3 / 2

- We will not write the final answer as 0.15 but

- We will carry out a step-by-step division as explained below.

Steps for step-by-step division

- Divide 3 by 2, to get a quotient (q) = 1 and remainder (r) = 1

- Write down the quotient in the answer line and the remainder just below it to its left as shown

$$0.3 / 2 = 0. \quad 1$$
$$1$$

- Subtract the quotient (1) from 9 and write it down

$$0.3 / 2 = 0. \overset{.}{1} \atop 1$$

Now, the next number for division is 18 and not 11.

On dividing 18 by 2, we get the q = 9 and r = 0. The result now looks as

$$0.3 / 2 = 0. \overset{.}{1} \quad \overset{8}{9} \atop 1 \quad 0$$

- Subtract the quotient (9) from 9 and write it down

$$0.3 / 2 = 0. \overset{.}{1} \quad \overset{8}{9} \quad \overset{.}{0} \atop 1 \quad 0$$

Now, the next number for division is 0 and not 9.

Repeat the division as explained before.

On dividing 0 by 2, we get q = 0 and r = 0. The result now looks as

$$0.3 / 2 = 0. \overset{.}{1} \quad \overset{8}{9} \quad \overset{.0}{0} \quad \overset{.9}{0} \atop 1 \quad 0 \quad 0$$

The final result upto 8 digits of accuracy is

$$0.3 / 2 = 0.\overset{.8}{1} \; \overset{.0}{9} \; \overset{.9}{0} \; \overset{.5}{4} \; \overset{.2}{7} \; \overset{.3}{6} \; \overset{.8}{1} \; \overset{.0}{9} \atop 1 \quad 0 \quad 0 \quad 1 \quad 1 \quad 0 \quad 1 \quad 0$$

and the result of 4/21 = 0.19047619

Example 2

Let us now see the details for 7 / 111

7/111 = 6/110 = 0.6/11

$$
\begin{array}{ccccc}
\overset{\bullet}{.9} & \overset{\bullet}{.3} & \overset{\bullet}{.6} & \overset{\bullet}{.9} & \overset{\bullet}{.3} \\
\end{array}
$$

$$
0.6 \; / \; 11 = 0. \; 0 \quad\quad 6 \quad\quad 3 \quad\quad 0 \quad\quad 6
$$
$$
 6 \quad\quad 3 \quad\quad 0 \quad\quad 6 \quad\quad 3
$$

Since the pattern is repeating, we can write down the answer as 7/111 = 0.063063063

Notes :

• For all numbers ending with 3, we can convert them to numbers ending with 1 by multiplying by 7.

e.g. if the divisor is 13, we can convert it to 91 by multiplying it by 7.

Hence,   3/13   =   21/91

5/23   =   35/161

• Similary, for numbers ending with 7, we can multiply by 3 to get a number ending with 1.

e.g. if the divisor is 17, we can convert it to 51 by multiplying it by 3.

Hence,   5/7   =   15/21

4/17   =   12/51

5/27   =   15/81.

Once the methods are clear for divisors ending with 9 and with 1, we can move over to other divisors ending with 8, 7 and 6.

40. *Examples for Auxiliary Fractions - Divisors ending with 1*

Sr No	Number	Converted to	Answer
1.	13/31 = 12/30	1.2/3	0.419355
2.	3/31 = 2/30	0.2/3	0.0967741
3.	1/21 = 0/20	0.0/2	0.047619
4.	2/21 = 1/20	0.1/2	0.095238
5.	5/21 = 4/20	0.4/2	0.2380952
6.	3/17 = 9/51 = 8/50	0.8/5	0.176471
7.	8/17 = 24/51 = 23/50	2.3/5	0.470588
8.	38/61 = 37/60	3.7/6	0.622951
9.	5/61 = 4/60	0.4/6	0.081967
10.	3/41 = 2/40	0.2/4	0.073171
11.	1/41 = 0/40	0.0/4	0.0243902
12.	5/51 = 4/50	0.4/5	0.0980392
13.	43/51 = 42/50	4.2/5	0.8431372
14.	72/81 = 71/80	7.1/8	0.888889
15.	16/91 = 15/90	1.5/9	0.1758241
16.	7/41 = 6/40	0.6/4	0.1707317
17.	23/81 = 22/80	2.2/8	0.2839506
18.	16/71 = 15/70	1.5/7	0.2253521
19.	2/51 = 1/50	0.1/5	0.0392156
20.	52/91 = 51/90	5.1/9	0.5714285

III) Divisors ending with '8'

Let us consider cases where the divisor ends with 8.

e.g. 18, 28, 48, 88 etc.

General Steps

• Add 2 to the denominator and use that as the divisor

• Remove the zero from the divisor and place a decimal point in the numerator at the appropriate position

• Carry out a step-by-step division as explained before

• Every division should return a quotient and a remainder and should not be carried out as a decimal division

• The quotient and the remainder is taken as the next base dividend as explained above

• Since the last digit of the divisor is 8 which has a difference of 1 from 9, we multiply the quotient by 1 and add to the base dividend at each stage to compute the gross dividend.

Example 1 :

Let us now see the details for 5 / 28

Steps

• Add 2 to the divisor to get 30

• The modified division is now 5 / 30

• Remove the zero from the denominator by placing a decimal point in the numerator giving the modified division as 0.5 / 3

• We will not write the final answer as 0.16666 but

• We will carry out a step-by-step division as explained below

Steps for step-by-step division

• Divide 5 by 3, to get a quotient (q) = 1 and remainder (r) = 2

• Write down the quotient in the answer line and the remainder just below it to its left as shown

$$0.5 / 3 = 0. \quad \underset{2}{1}$$

Now, the base dividend is 21

Add the quotient digit (1) to the initial dividend to get 22.

The next number for division is 22 which on division by 3 gives q = 7 and r = 1. The result now looks as

$$0.5 / 3 = \quad 0. \quad \underset{2}{1} \quad \underset{1}{7}$$

Now, the base dividend is 17.

Add the quotient digit (7) to the base dividend to get 24.

The next number for division is 24 which on division by 3 gives q = 8 and r = 0. The result now looks as

$$0.5 / 3 = 0. \quad \underset{2}{1} \quad \underset{1}{7} \quad \underset{0}{8}$$

Repeat the division to get the final answer as shown below.

The final result upto 5 digits of accuracy is

$$0.5 / 3 = 0. \quad \underset{2}{1} \quad \underset{1}{7} \quad \underset{0}{8} \quad \underset{1}{5} \quad \underset{2}{6}$$

and the result of 5/28 = 0.17856.

41. Examples for Auxiliary Fractions - Divisors ending with 8

Sr No	Number	Converted to	Answer
1.	1/18 = 1/20	0.1/2	0.0555555
2.	5/18 = 5/20	9.5/2	0.2777777
3.	1/28 = 1/30	0.1/3	0.0357142

4.	5/26 = 15/78 = 15/80	1.5/8	0.1923076
5.	6/28 = 6/30	0.6/3	0.2142857
6.	71/88 = 71/90	7.1/9	0.8068181
7.	1/48 = 1/50	0.1/5	0.0208333
8.	21/48 = 21/50	2.1/5	0.4375
9.	2/38 = 2/40	0.2/4	0.0526315
10.	3/14 = 6/28 = 6/30	0.6/3	0.2142857
11.	5/58 = 5/60	0.5/6	0.0862068
12.	7/78 = 7/80	0.7/8	0.0897435
13.	1/17 = 4/68 = 4/70	0.4/7	0.0588235
14.	5/88 = 5/90	0.5/9	0.0568181
15.	43/98 = 43/100	4.3/10	0.4387755
16.	7/58 = 7/60	0.7/6	0.1206896
17.	15/78 = 15/80	1.5/8	0.1923076
18.	4/38 = 4/40	0.4/4	0.1052631
19.	31/98 = 31/100	3.1/10	0.3163265
20.	3/48 = 3/50	0.3/5	0.0625

IV) Divisors ending with '7'

Let us consider cases where the divisor ends with 7.
e.g. 17, 27, 37, 57 etc.

General Steps

* Add 3 to the denominator and use that as the divisor
* Remove the zero from the divisor and place a decimal point in the numerator at the appropriate position
* Carry out a step-by-step division as explained before
* Every division should return a quotient and a remainder and

should not be a decimal division

- The quotient and the remainder is taken as the next base dividend as explained before

- Since the last digit of the divisor is 7 which has a difference of 2 from 9, we multiply the quotient by 2 and add to the base dividend at each stage to compute the next gross dividend.

Example 1 :

Let us now see the details for 5 / 27.

Steps

- Add 3 to the divisor to get 30

- The modified division is now 5 / 30

- Remove the zero from the denominator by placing a decimal point in the numerator giving the modified division as 0.5 / 3

- We will not write the final answer as 0.16666 but

- We will carry out a step-by-step division explained below.

Steps for step-by-step division

- Divide 5 by 3, to get a quotient (q) = 1 and remainder (r) = 2

- Write down the quotient in the answer line and the remainder just below it to its left as shown

$$0.5 / 3 = 0. \quad 1$$
$$ 2$$

- Now, the initial base dividend is 21

- Multiply the quotient digit (1) by 2 and add to the base dividend 21 to get 23

- The next number for division is 23 which on division by 3 gives q = 7 and r = 2. The result now looks as

$$0.5 / 3 = 0. \quad \underset{2}{1} \quad \underset{2}{7}$$

- Now, the base dividend is 27

- Multiply the quotient digit (7) by 2 to get 14 and add it to the base dividend (27) to get 41

- The next number for division is 41 which on division by 3 gives q = 13 and r = 2. The result now looks as

$$0.5 / 3 = 0. \quad \underset{2}{1} \quad \underset{2}{7} \quad \underset{2}{(13)}$$

- At this point, note that the quotient digit has become a 2-digit number viz 13. Like before, it is multiplied by 2 to get 26 which is added as shown below :

$$\begin{array}{r} 13 \\ 2 \\ \underline{26} \\ 59 \end{array}$$

- The remainder '2' is added to the tens digit of 13 and 26.

- Now, we divide 59 by 3 to get the quotient as 19 and the remainder as 2. The working appears as shown below:

$$0.5/3 = 0. \quad \underset{2}{1} \quad \underset{2}{7} \quad \underset{2}{(13)} \quad \underset{2}{(19)}$$

- 19 is multiplied by 2 to get 38, which is added as shown below :

$$\begin{array}{r} 19 \\ 2 \\ 38 \\ \hline 77 \end{array}$$

- Now, we divide 77 by 3 to get the quotient as 25 and the remainder as 2. The working appears as shown below:

$$0.5/3 = \quad 0. \quad \underset{2}{1} \quad \underset{2}{7} \quad \underset{2}{(13)} \quad \underset{2}{(19)} \quad \underset{2}{(25)}$$

- 25 is multiplied by 2 to get 50, which is added as shown below :

$$\begin{array}{r} 25 \\ 2 \\ 50 \\ \hline 95 \end{array}$$

- On dividing 95, we get the quotient as 31 and the remainder as 2. The working appears as shown below:

$$0.5/3 = \quad 0. \quad \underset{2}{1} \quad \underset{2}{7} \quad \underset{2}{(13)} \quad \underset{2}{(19)} \quad \underset{2}{(25)} \quad \underset{2}{(31)}$$

- We will now see how to build the final result, by retaining a single digit at each location. The answer consists of the following numbers :

$$0. \quad 1 \quad 7 \quad (13) \quad (19) \quad (25) \quad (31)$$

- Start from the right, retain 1, carry 3 to the left to get

$$0. \quad 1 \quad 7 \quad (13) \quad (19) \quad (28) \quad 1$$

- Retain 8 from 28, carry 2 to the left to get

$$0. \quad 1 \quad 7 \quad (13) \quad (21) \quad 8 \quad 1$$

- Repeat to get the final result as

0.	1	7	(15)	1	8	1
0.	1	8	5	1	8	1

- The final result with 6 digits of accuracy is
  $5/27 = 0.185181$.

42. Examples for Auxiliary Fractions - Divisors ending with 7

Sr No	Number	Converted to	Answer
1.	1/17 = 1/20	0.1/2	0.058823529
2.	4/17 = 4/20	0.4/2	0.235294118
3.	5/27 = 5/30	0.5/3	0.185185185
4.	19/27 = 19/30	1.9/3	0.703703703
5.	2/37 = 2/40	0.2/4	0.054054054
6.	73/87 = 73/90	7.3/9	0.83908046
7.	1/47 = 1/50	0.1/5	0.021276596
8.	4/47 = 4/50	0.4/5	0.0851063
9.	34/37 = 34/40	3.4/4	0.918918919
10.	3/67 = 3/70	0.3/7	0.044776119
11.	5/57 = 5/60	0.5/6	0.087719298
12.	69/77 = 69/80	6.9/8	0.896103896
13.	14/57 = 14/60	1.4/6	0.245614
14.	5/87 = 5/90	0.5/9	0.0574712
15.	43/97 = 43/100	4.3/10	0.4432989
16.	23/77 = 23/80	2.3/8	0.2987012
17.	3/47 = 3/50	0.3/5	0.0638297
18.	41/97 = 41/100	4.1/10	0.4226804
19.	15/67 = 15/70	1.5/7	0.2238805
20.	7/87 = 7/90	0.7/9	0.0804597

IV) Numbers ending with '6'

Let us consider cases where the divisor ends with 6.
e.g. 26, 36, 56, 86 etc.

General Steps

- Add 4 to the denominator and use that as the divisor

- Remove the zero from the divisor and place a decimal point in the numerator at the appropriate position

- Carry out a step-by-step division as explained before

- Every division should return a quotient and a remainder and should not be a decimal division

- The quotient and the remainder is taken as the next base dividend as explained before

- Since the last digit of the divisor is 6 which has a difference of 3 from 9, at each stage, we multiply the quotient by 3 and add to the base dividend to compute the next gross dividend.

Example 1 :

Let us now see the details for 6 / 76.

Steps

- Add 4 to the divisor to get 80

- The modified division is now 6 / 80

- Remove the zero from the denominator by placing a decimal point in the numerator giving the modified division as 0.6 / 8

- We will not write the final answer as 0.075 but

- We will now carry out a step-by-step division as explained below.

Steps for step-by-step division

- Divide 6 by 8, to get a quotient (q) = 0 and remainder (r) = 6

- Write down the quotient in the answer line and the remainder just below it to its left as shown

$$0.6 / 8 = 0. \quad 0$$
$$6$$

- Now, the initial base dividend is 60

- Multiply the quotient digit (0) by 3 and add to the base dividend 60 to get 60. This is a redundant step here since the quotient digit is zero, but it is shown for uniformity.

- The next number for division is 60 which on division by 8 gives q = 7 and r = 4. The result now looks as

$$0.6 / 8 = 0. \quad 0 \quad 7$$
$$6 \quad 4$$

- Now, the base dividend is 47

- Multiply the quotient digit (7) by 3 to get 21 and add it to the base dividend (47) to get 68

- The next number for division is 68 which on division by 8 gives q = 8 and r = 4. The result now looks as

$$0.6 / 8 = 0. \quad 0 \quad 7 \quad 8$$
$$6 \quad 4 \quad 4$$

- Repeat the division to get the final answer as shown below.

- The final result upto 8 digits of accuracy is

$$0.6/8 = 0. \quad 0 \quad 7 \quad 8 \quad 9 \quad 4 \quad 7 \quad 3 \quad 8$$
$$6 \quad 4 \quad 4 \quad 0 \quad 4 \quad 0 \quad 4 \quad 4$$

and the result of 6/76 = 0.07894738.

43. Examples for Auxiliary Fractions - Divisors ending with 6

Sr No	Number	Converted to	Answer
1.	7/26 = 7/30	0.7/3	0.2692307
2.	13/46 = 13/50	1.3/5	0.282608696
3.	25/56 = 25/60	2.5/6	0.446428571
4.	21/76 = 21/80	2.1/8	0.276315789
5.	7/36 = 7/40	0.7/4	0.194444444
6.	73/86 = 73/90	7.3/9	0.848837209
7.	1/46 = 1/50	0.1/5	0.02173913
8.	39/76 = 39/80	3.9/8	0.513157895
9.	13/66 = 13/70	1.3/7	0.196969697
10.	5/36 = 5/40	0.5/4	0.138888889
11.	43/46 = 43/50	4.3/5	0.934782609
12.	17/56 = 17/60	1.7/6	0.303571429
13.	59/86 = 59/90	0.5/9	0.686046512
14.	43/76 = 43/80	4.3/8	0.565789474
15.	27/66 = 27/70	2.7/7	0.409090909
16.	5/56 = 5/60	0.5/6	0.892857
17.	13/76 = 13/80	1.3/8	0.1710526
18.	11/96 = 11/100	1.1/10	0.1145833
19.	23/86 = 23/90	2.3/9	0.2674418
20.	7/36 = 7/40	0.7/4	0.1944444

VI. Other Divisors

In the sections given above, we have seen divisors ending with 1, 6, 7, 8 and 9. What do we do with divisors ending with 2, 3, 4 and 5? We can convert such divisors, by a suitable multiplication, to divisors for which we have seen vedic techniques. Use the table given below for guidance.

Sr No	Divisor ending with	Multiply By	Example	After multiplication
1.	2	5	13/42	65/210
		4	17/22	68/88
2.	3	3	13/43	39/129
		7	13/43	91/301
3.	4	5	13/34	65/170
		7	13/34	91/238
4.	5	2	13/235	26/470
5.	6	3	5/26	15/78
		5	5/26	25/130

We can see from this table that if a proper digit is used to multiply the numerator and the denominator, we can obtain a fraction wherein we can apply one of the known techniques and derive the answer quickly. Sometimes, we may have to carry out the multiplication twice to get it in a standard form.

44. Examples for Auxiliary Fractions - Miscellaneous Exercises

Sr No	Number	Converted to	Answer
1.	5/13 = 15/39	1.5/4	0.384615385
2.	7/32 = 35/160	3.5/16	0.21875
3.	5/28 = 5/30	0.5/3	0.17857142
4.	21/44 = 42/88	4.2/9	0.477272727
5.	53/82 = 265/410	26.5/41	0.646341463
6.	7/23 = 21/69	2.1/7	0.304347826

7.	8/43 = 24/129	2.4/13	0.186046512
8.	5/14 = 10/28	1.0/3	0.357142857
9.	37/42 = 185/210	18.5/21	0.880952381
10.	2/63 = 6/189	0.6/19	0.031746032
11.	7/72 = 35/360	3.5/36	0.097222222
12.	9/23 = 27/69	2.7/7	0.391304348
13.	12/53 = 36/159	3.6/16	0.226415094
14.	19/24 = 38/48	3.8/5	0.791666667
15.	11/34 = 22/68	2.2/7	0.323529412
16.	13/34 = 26/68 = 26/70	2.6/7	0.382352941
17.	7/13 = 21/39 = 21/40	2.1/4	0.538461538
18.	9/26 = 9/30	0.9/3	0.346153846
19.	27/53 = 81/159 = 81/160	8.1/16	0.509433962
20.	8/63 = 24/189 = 24/190	2.4/19	0.126984126

CHAPTER **12**

Mishrank (Vinculum)

The technique of mishrank is very powerful and provides us with a method to convert digits in a number, which are greater than 5, to digits less than 5. After the conversion, all arithmetic operations are carried out using the converted number, which make the operation very simple and fast.

It is not necessary to convert all the digits but a judicious conversion of certain digits (above 5) can decrease the computation effort considerably. The reader has to be patient while learning this totally new and fascinating technique and gain expertise in order to use this method effectively to great advantage.

We will begin by seeing the method to convert any given number to its mishrank equivalent and back to its original value.

a) Conversion to Mishrank

Mishrank digits are computed for those digits which are greater than 5.

The steps involved are :

- Subtract the given digit from 10 and write it with a bar above it.
- Add one to the digit on the left.

Examples

Sr No	Given Number	After Conversion to Mishrank	Explanation
1.	1 4 8	1 5 $\overline{2}$	Subtract 8 from 10, to get 2, add 1 to 4 to get 5
2.	2 2 8 5	2 3 $\overline{2}$ 5	
3.	2 2 8 8 4	2 3 $\overline{1}$ $\overline{2}$ 4	Subtract 8 in tens place from 10 to get 2, add 1 to the 8 to its left to get 9. Repeat for 9, subtract it from 10 to get 1, add 1 to 2 to its left.
4.	7 6 3 2	1 $\overline{2}$ $\overline{4}$ 3 2	Subtract 6 from 10 tlo get 4, add 1 to left to get 8, subtract it from 10 to get 2, add 1 to left, which is zero to get 1 in left-most place.
5.	2 3 8 4 9	2 4 $\overline{2}$ 5 $\overline{1}$	
6.	8 8 8 8	1 $\overline{1}$ $\overline{1}$ $\overline{1}$ $\overline{2}$	
7.	2 0 7 4 5	2 1 $\overline{3}$ 4 5	

The mishrank digit has a negative value in the number.

Thus, in the first example above,

1 5 $\overline{2}$ is equivalent to 150 – 2 i.e. 148

Similarly, 2 3 $\overline{2}$ 5 is equivalent to 2305 – 20 = 2285 while
1 $\overline{1}$ $\overline{1}$ $\overline{1}$ $\overline{2}$ is equivalent to 10000 – 1112 = 8888

b) Conversion from Mishrank

Only mishrank digits, which are marked with a bar, are re-converted back to their original values, as follows :

• Subtract 1 from the non-mishrank digit to the immediate left of the mishrank digit and write it down.

• Subtract the mishrank digit from 10 and write it down.

• Repeat for all mishrank digits.

Sr No	Given Number	On Conversion to Original Number
1.	1 5 $\overline{4}$	1 4 6
2.	2 2 $\overline{3}$ 5	2 1 7 5
3.	2 1 $\overline{4}$ 5 5	2 0 6 5 5
4.	2 $\overline{3}$ $\overline{3}$ $\overline{3}$	1 6 6 4
5.	5 0 $\overline{3}$ 4 5	4 9 7 4 5
6.	2 5 $\overline{3}$ 5 $\overline{2}$	2 4 7 4 8
7.	3 $\overline{4}$ 0 $\overline{2}$ 3	2 5 9 8 3

In example 4, we have a series of mishrank digits and we can use the sutra 'All from 9 and last from 10' to convert to the original number.

Mishrank can be used very effectively in operations like addition, subtraction, multiplication, division, finding squares and cubes of numbers. We will see some examples to see its tremendous power.

c) Application in multiple addition and subtraction

Let us consider the following problem :
232 + 4151 − 2889 + 1371

This could either be solved by first adding 232 and 4151, then subtracting 2889 from the sum and then adding 1371 to the result.

Or, we could add 232, 4151 and 1371 and then subtract 2889 from the sum.

Instead, we could speed up the operation by converting the subtraction to an addition operation as shown below :

Let us write − 2889 as + $\overline{2}\ \overline{8}\ \overline{8}\ \overline{9}$

Here, each digit is written as a negative digit.

Now, the problem can be rewritten as

232 + 4151 + $\overline{2}\ \overline{8}\ \overline{8}\ \overline{9}$ + 1371
or written vertically, it appears as

$$
\begin{array}{r}
2\ 3\ 2 \\
+\quad 4\ 1\ 5\ 1 \\
+\quad \overline{2}\ \overline{8}\ \overline{8}\ \overline{9} \\
+\quad 1\ 3\ 7\ 1 \\
\hline
3\ \overline{2}\ 7\ \overline{5}
\end{array}
$$

Since, this is a single operation, it is faster and causes less strain to the mind. The mishrank digits in the result can be converted to normal digits to obtain the final answer as 2865.

d) Application in subtraction

Example 1 : Consider the subtraction of 6869 from 8988

The subtraction operation would be slow and mentally tiring, with a carry at almost every step. The operation can be simplified and speeded up by using either of the following techniques :

- Convert the subtraction to addition of a number as shown in the previous example.

- Carry out a digit-by-digit subtraction, without any carry

Let's see the *first* technique.

6869 is written with a bar on top of each digit to signify that it is a negative digit.

$$
\begin{array}{r}
8 \quad 9 \quad 8 \quad 8 \\
+ \quad \overline{6} \quad \overline{8} \quad \overline{6} \quad \overline{9} \\
\hline
2 \quad 1 \quad 2 \quad \overline{1} \\
\hline
\end{array}
$$

which is equal to 2119. Here, in the unit's place, $8 + \overline{9}$ gives $\overline{1}$.

Let's see the *second* technique.

The same result can be obtained if we simply subtract each digit without a carry.

Let's see the operation :

$$
\begin{array}{r}
8 \quad 9 \quad 8 \quad 8 \\
- \quad 6 \quad 8 \quad 6 \quad 9 \\
\hline
2 \quad 1 \quad 2 \quad \overline{1} \\
\hline
\end{array}
$$

Here, in the unit's place, $8 - 9$ gives $\overline{1}$, which gives 2119 on conversion from mishrank.

e) Application in multiplication

Example 1 : Consider the multiplication of 69 with 48.

The operation can be carried out by the cross-multiplication technique explained in Chapter 5. The computation can be simplified further if we convert the non-mishrank digits to mishrank digits. We will see the method below.

$$6\ 9\ =\ 7\ \overline{1}$$
$$4\ 8\ =\ 5\ \overline{2}$$

Now, we will carry out a cross multiplication of these two numbers.

$$
\begin{array}{ccc}
 & 7 & \overline{1} \\
\times & 5 & \overline{2} \\
\hline
\end{array}
$$

$$35\ :\ \overline{1}\ \overline{9}\ :\ 2$$

$$=\quad 34 : \overline{9} : 2 \text{ (Carry } \overline{1} \text{ to the left and add to 35 to}$$
$$\text{get 34)}$$

$$=\quad 33 : 1 : 2$$

$$=\quad 3312$$

Example 2 : Consider the multiplication of 882 by 297

The corresponding mishrank digits are

$8\ 8\ 2\ =\ 9\ \overline{2}\ 2$ (We convert the 8 in the ten's place and do not convert the 8 in the hundreds place)

$2\ 9\ 7\ =\ 3\ 0\ \overline{3}$

Now, we will carry out a cross multiplication of these two numbers.

$$
\begin{array}{r}
9 \quad \bar{2} \quad 2 \\
\times \quad 3 \quad 0 \quad \bar{3} \\
\hline
27 : \bar{6} : \bar{2}\bar{1} : 6 : \bar{6}
\end{array}
$$

$$
\begin{aligned}
&= \quad 27 : \bar{8} : \bar{1} : 6 : \bar{6} \\
&= \quad 26 : 1 : 9 : 5 : 4 \\
&= \quad 261954
\end{aligned}
$$

f) Application in division

Example 1 : Consider the division of 25382 by 77

77 can be written as $8\bar{3}$, where we convert 7 in unit's place to its mishrank digit.

$$
\begin{array}{c|ccccccc}
8^{\bar{3}} & \underline{25} & 3 & 8 : 2 & 0 & 0 & 0 \\
 & & 1 & 6 \quad 2 & 1 & 4 & 1 \\
 & & -(-9) & -(-6)\;-(-27) & -(-18) & -(-9) & -(-18) \\
\hline
 & & 22 & 74 \quad 49 & 28 & 49 & 28 \\
\hline
 & 3 & 2 & 9 : 6 & 3 & 6 & 3
\end{array}
$$

Division is carried out by the same technique as explained before. It will be seen that there is no need to make any adjustment at any column and the procedure can progress very fast.

The final result is 329 with remainder 49 or in decimal form, it is 329.6363.

g) Application in squares

In order to get the square of any number, we need to compute duplex values of a lot of numbers. If some of the digits are above 5, the computation of the duplex becomes tedious. Consider the square of 897. The computation of the duplex is quite time-consuming since the digits are big. We will see the magic of the mishrank digits now.

897 = $90\overline{3}$
      = $1\overline{1}0\overline{3}$

Both are valid mishrank equivalent numbers.

We will now compare the computation of the square of the given number for all these numbers.

897^2 = 64/144/193/126/49 = 804609
$90\overline{3}^2$ = 81/0/$5\overline{4}$/0/9 = 8 1 $\overline{5}\,\overline{4}$ 0 9 = 804609.

Let's look at the next conversion, i.e. $1\overline{1}0\overline{3}$

$1\overline{1}0\overline{3}^2$ = 1/$\overline{2}$/1/$\overline{6}$/6/0/9 = $8\overline{1}6609$ = 804609 as before.

We can see that the computations become increasingly simpler.

h) Application in cubes

Consider the cube of 89 where 89^3 = 704969

The computation without using the mishrank would appear as follows :

89^3	=	512	576	648	729
			1152	1296	
		704	9	6	9

The computation is very tedious and cumbersome.

Let's convert to mishrank and see what is the result.

89 = $9\overline{1}$

$$9\overline{1}^3 = 729 \quad \begin{array}{cccc} & \overline{8}\,\overline{1} & 9 & \overline{1} \\ & \overline{1}\,\overline{6}\,\overline{2} & 18 & \end{array}$$

$$729 : \overline{2}\,\overline{4}\,\overline{3} : 27 : \overline{1}$$

$$= 705 : \overline{1} \quad : 7 : \overline{1}$$
$$= 704969$$

This is easier and faster.

In each of the operations, it can be seen that the operation on the corresponding mishrank number is both easy and fast. The reader has to practice to convert numbers to their mishrank equivalents and back again, to fully avail of the power of this technique.

45. *Examples in use of Mishrank*

Sr No	Example	Answer
1.	7988 + 1786 − 1233 − 2146	6395
2.	19.56 − 67.87 + 72.35	24.04
3.	2352 − 1779	573
4.	34.23 − 2.69	31.54
5.	69 × 88	6072
6.	97 × 38	3686
7.	189 × 278	52542
8.	77.7 × 5.78	449.106
9.	78.94 × 69.97	5523.4318
10.	779²	606841
11.	47.8²	2284.84
12.	7251 / 49	147.97959
13.	1279 / 59	21.677966
14.	49492 / 572	86.52447
15.	29³	24389

CHAPTER 13

Simultaneous Equations

Everyone is familiar with simultaneous equations wherein we have to solve two equations in two unknowns, say x and y.

The normal method consists of multiplying each equation by a suitable number so that the co-efficients of either x or y become same in both the equations, which can then be subtracted out of the equations, leaving a single equation in one unknown.

Vedic maths provides a simple method to solve the equations without going through this lengthy process and gives the values of x and y in a single line. This would be similar to 'Cramer's Rule' but is easier to remember and use.

We will consider pairs of equations each with two unknowns, say x and y.

$$\text{Example} \quad 4x + 7y = 5 \quad \dots (1)$$
$$3x + 9y = 10 \quad \dots (2)$$

The normal method to solve such simultaneous equations is to eliminate any one unknown by suitable multiplication of each equation by a constant and then solving for the other unknown variable. In the above example, multiply equation (1) by 3 and equation (2) by 4 and subtract.

$$12x + 21y = 15 \quad \dots (1)$$
$$12x + 36y = 40 \quad \dots (2)$$
$$\overline{-15y = -25}$$

Which gives $y = 5 / 3$.

On substituting this value of y in equation (1), we get

$$4x + 7 * 5 / 3 = 5$$

Giving $4x = -20 / 3$

∴ $x = -5 / 3$

We will now see the techniques provided by Vedic maths.

Equations in 2 unknowns can be divided into 3 categories and we will see the techniques for each of them.

- When the ratio of the coefficients of either x or y is the same as that of the constants

- When the ratio of the coefficients of x and y are interchanged

- All other types not falling in these two categories

Category 1

Consider the example :

$$3x + 4y = 10$$
$$5x + 8y = 20$$

In this example, the ratio of the coefficients of y is $4/8 = 1/2$. The ratio of the constants is $10/20 = 1/2$.

Since both are equal, the value of x in such a case is equal to 0. If the co-efficients of x were in the same ratio, value of y would be zero.

The value of y can be obtained from the first equation by substituting $x = 0$, giving the equation as

$$4y = 10$$
or $y = 5/2$

Category 2

Now, let us consider the method to solve simultaneous equations where the coefficients of x and y are interchanged. Let us consider the example given below.

$$45 x - 23 y = 113 \quad \dots \quad (1)$$
$$23 x - 45 y = 91 \quad \dots \quad (2)$$

We will add and then subtract the given equations to get a simplified set as shown below

$$68 x - 68 y = 204 \quad \dots \quad (3)$$
$$22 x + 22 y = 22 \quad \dots \quad (4)$$

Dividing eq (3) by 68 and eq (4) by 22, we get

$$x \quad - y \quad = 3 \quad \dots \quad (5)$$
$$x \quad + y \quad = 1 \quad \dots \quad (6)$$

This is a trivial set, which gives the values of
x = 2 and y = −1.

Category 3

Let us consider the method to solve any general simultaneous equation. Let us consider the example solved at the beginning of this chapter.

$$4 x + 7 y = 5 \quad \dots \quad (1)$$
$$3 x + 9 y = 10 \quad \dots \quad (2)$$

Let us represent this as a set of coefficients as follows:

$$a \quad b \quad c$$
$$p \quad q \quad r$$

where a = 4, b = 7, c = 5, p = 3, q = 9, r = 10.

We start with the coefficient at the center in the upper row viz 'b' and get the value of 'x' by using the formula

$$x = \frac{b * r - c * q}{b * p - a * q}$$

Notice the cross multiplication which is used extensively here in this formula.

b * r and c * q are cross-multiplications, from the centre on the right.

Similarly, b * p and a * q are cross multiplications, from the centre on the left.

To get the value of y,

$$y = \frac{c * p - a * r}{b * p - a * q}$$

Here, again, c * p and a * r are cross-multiplications while the denominator is the same as before.

So, we get x $= \dfrac{7 * 10 - 5 * 9}{7 * 3 - 4 * 9}$

$= \dfrac{70 - 45}{21 - 36} = 25 / (-15) = -(5/3)$

and y $= \dfrac{5 * 3 - 4 * 10}{(-15)} = \dfrac{15 - 40}{(-15)}$

$= -25 / (-15)$

$= 5/3$

46. Exercises for practice

a) $2x + 3y = 8$
    $4x + 5y = 14$ ... x =1, y =2

b) $x - y = 7$
    $5x + 2y = 42$ .... x=8, y=1

c) $2x + y = 5$
    $3x - 4y = 2$ ... x = 2, y=1

d) $5x - 3y = 11$
    $6x - 5y = 9$ ... x=4, y=3

e) $11x + 6y = 28$
    $7x - 4\ y = 10$ ... x=2, y=1

f) $12\ x + 8y = 7$
    $16\ x + 16y = 14$ ... x=0, y=7/8

g) $6x + 7y = 8$
    $19x + 14\ y = 16$ ... x=0, y=8/7

h) $9x + 2y = 2$
    $3x + 5y = 5$ ... x=0, y=1

i) $37\ x + 29\ y = 95$
    $29x + 37\ y = 103$ ...x=1, y=2

j) $12x + 17y = 53$
    $17x + 12y = 63$ ... x=3, y=1

CHAPTER 14

Osculators

The concept of 'Osculator' is useful to check the divisibility of a given number by divisors ending with 9 or 1 or a multiple thereof.

e,g. Check the divisibility of 52813 by 59 or
of 32521 by 31 or
of 35213 by 13 or 17.

An osculator is a number defined for a number ending in 9 or 1 and is obtained from the number by a simple mechanism described in this chapter.

The use of osculators would be severely limited if only these two categories of numbers, viz ending with 9 or 1 are considered. Interestingly, numbers ending with 3 and 7 can also be converted to numbers ending with 1 or 9 by a suitable multiplication. The same technique can be used for all such numbers too.

Osculators are categorized into two main types, viz positive and negative, depending on whether the number ends with 9 or with 1.

Positive Osculators

The osculator of a number ending in 9 is considered positive.

It is obtained by

- Dropping 9
- Adding 1 to the remaining number

The osculator for 19 is 2, for 29 it is 3 and for 59, it is 6.

If the given number does not end in 9 but in 1 or 3 or 7, we can multiply it by 9, 3 or 7 respectively, to convert it to a number which ends in 9 and then compute the osculator by dropping the 9.

Examples are given below of how to compute the osculators for various numbers.

Once the osculator of a number has been computed, it is a simple matter to check the divisibility of any given number by it.

Example : Let us check the divisibility of 228 by 19.

Steps

- Compute the osculator of the divisor (19) which is 2.

Digit 1

- Start with the computation from the right hand digit (8) of the dividend

- Multiply it by the osculator to get 16 (8 x 2)

Digit 2

- Add it to the digit to its left to get 18 (16 + 2)

- Multiply it by the osculator to get 36 (18 x 2)

- Subtract maximum possible multiples of the divisor from it. Hence, 19 can be subtracted from it to get 17.

Digit 3

• Add it to the digit to its left to get 19 (17 + 2)

Since 19 is divisible by 19, the given number is also divisible by 19.

Computation of Positive Osculators for different numbers

Given Number ends with	Operation	Given Number	Osculator
9	Drop 9 and add 1 to previous digit	9 19 29	1 2 3
7	Multiply given number by 7	7 → 49 17 → 119	5 12
3	Multiply given number by 3	13 → 39 23 → 69	4 7
1	Multiply given number by 9	21 → 189 11 → 99	19 10

47. Examples for checking the divisibility of the following by 19

Sr No	Given Number	Divisibility
1.	21	No
2.	57	Yes
3.	131	No
4.	225	No
5.	437	Yes
6.	2337	Yes

7.	2774	Yes
8.	1015949	Yes
9.	11021232	No
10.	13201999	No
11.	665	Yes
12.	11628	Yes
13.	48412	Yes
14.	1765489	No
15.	2100897	No
16.	10162915	No
17.	4773	No
18.	2169	No
19.	4655	Yes
20.	6783	Yes

48. Examples for checking the divisibility of the following by 29

Sr No	Given Number	Divisibility
1.	62	No
2.	87	Yes
3.	242	No
4.	232	Yes
5.	319	Yes
6.	522	Yes
7.	2119	No
8.	32896	No
9.	93148	No
10.	2434521	Yes

11.	3420675	No
12.	87519	No
13.	45936	Yes
14.	1586909	Yes
15.	137779	Yes
16.	622224	Yes
17.	16598632	No
18.	59678	No
19.	149292	Yes
20.	819	No

49. Examples for checking the divisibility of the following by 13

Sr No	Given Number	Divisibility
1.	52	Yes
2.	78	Yes
3.	143	Yes
4.	162	No
5.	14061	No
6.	42705	Yes
7.	33054	No
8.	273273	Yes
9.	6877	Yes
10.	59709	Yes
11.	438709	No
12.	269017	No
13.	995683	Yes
14.	18876	Yes

15.	80369	No
16.	428531	No
17.	326690	Yes
18.	926159	Yes
19.	8133	No
20.	14631	No

50. Examples for checking Divisibility - Miscellaneous examples

Sr No	Given Number	Divide by	Divisibility
1.	161928	39	Yes
2.	5332	49	No
3.	10385	49	No
4.	59383	49	No
5.	1092	21	Yes
6.	27962	11	Yes
7.	21953	53	No
8.	4914	39	Yes
9.	33803	79	No
10.	33733	79	Yes
11.	41019	33	Yes
12.	11203	29	No
13.	21098	31	No
14.	14705	19	No
15.	145245	69	Yes
16.	10891	29	No
17.	36429	21	No
18.	2019	19	No
19.	21217	49	Yes
20.	22149	69	Yes

Negative Osculators

Let us now consider negative osculators, which are defined for numbers ending with 1.

To get the osculator of a number ending in 1

- Drop 1
- The remaining number is the osculator

The osculator for 11 is 1, for 21 it is 2 and for 61, it is 6.

Given Number ends with	Operation	Given Number	Osculator
9	Multiply given number by 9	9 → 81 19 → 171	8 17
7	Multiply given number by 3	7 → 21 17 → 51	2 5
3	Multiply given number by 7	13 → 91 23 → 161	9 16
1	Just drop 1	21 31	2 3

Example : Let us work out the divisibility of 2793 by 21.

Steps

- Start with the osculator of the divisor (21) which is 2
- Mark all alternate digits starting from the digit in the tens place as negative by placing a bar on top

  $\bar{2}$ 7 $\bar{9}$ 3

Digit 1

- Start with the computation from the right hand digit (3) of the dividend

- Multiply it by the osculator to get 6 (3 x 2)

Digit 2

- Add it to the digit to its left to get $\overline{3}$ (6 + $\overline{9}$)
- Multiply it by the osculator to get $\overline{6}$ ($\overline{3}$ x 2)
- Subtract maximum possible multiples of the divisor from it. We cannot subtract anything here.

Digit 3

- Add it to the previous digit to get 1 (7 + $\overline{6}$)
- Multiply it by the osculator to get 2 (1 × 2)

Digit 4

- Add it to the previous digit to get 0 ($\overline{2}$ + 2)
- Multiply it by the osculator to get 0 (0 x 2)

Since 0 is divisible by 21, the given number is also divisible by 21.

51. Examples for checking the divisibility by 21

Sr No	Given Number	Divisibility
1.	318	No
2.	315	Yes
3.	25914	Yes
4.	5377	No
5.	53.403	Yes
6.	132.71	No
7.	29883	Yes
8.	67221	Yes
9.	1563.91	No
10.	41075	No
11.	24356	No

12.	452823	Yes
13.	449505	Yes
14.	2528.61	Yes
15.	44186.52	Yes
16.	4557	Yes
17.	15603	Yes
18.	337	No
19.	1546	No
20.	34712	No

52. Examples for checking the divisibility by 31

Sr No	Given Number	Divisibility
1.	6603	Yes
2.	6604	No
3.	44020	Yes
4.	52793	Yes
5.	404395	Yes
6.	126307	No
7.	15349	No
8.	624061	Yes
9.	3763.09	Yes
10.	53227	Yes
11.	21063	No
12.	20134	No
13.	99231	Yes
14.	35061	Yes
15.	41209	No
16.	77321	No

17.	3295	No
18.	57474	Yes
19.	45663	Yes
20.	2134	No

53. Miscellaneous examples

Sr No	Given Number	Divide by	Divisibility
1.	11235	41	No
2.	11234	41	Yes
3.	18733	41	No
4.	12213054	61	Yes
5.	12954	51	Yes
6.	1185393	51	Yes
7.	31682	31	Yes
8.	111888	31	No
9.	51112	21	No
10.	216993	21	Yes
11.	11775	41	No
12.	10223	51	No
13.	210661	11	Yes
14.	179456	41	No
15.	179445	61	No
16.	113465	11	Yes
17.	213111	21	No
18.	101989	41	No
19.	11669	31	No
20.	158661	51	Yes

CHAPTER 15

Applications of Vedic Maths

We now come to the most interesting chapter of this book, in which we will learn how to apply the techniques learnt so far. After reading and solving the problems in this chapter, the reader will realize the immense benefit that is derived by using the simple and elegant methods of vedic maths in solving a variety of problems.

I have divided this chapter into 3 parts. In the first part, I have listed twenty questions selected from various competitive exams. The reader is encouraged to solve these problems independently and track the time taken for solving. This part is followed by the solutions to the problems, using techniques from vedic maths, including a reference to the relevant chapter. The solutions are brief but sufficient to follow and the reader will easily grasp the application of the techniques learnt so far. The ease and the efficiency would be appreciated. In the third part, I have included some more problems for practice which the reader should solve and gain expertise in the subject.

Sample Problems

1) Which is the greatest 6-digit number which is a perfect square?

2) If the length and the breadth of a square is increased and reduced by 5% respectively, what is the percentage increase or decrease in the area?

3) Compute $896 \times 896 - 204 \times 204$

4) If all the digits in numbers 1 to 24 are written down, would the number so formed be divisible by 9?

5) What should be the value of * in the following number so that it is divisible by 11?
   - 8287*845
   - *381

6) Which of the following numbers is divisible by 99?
   135792,   114345,   3572404,   913464

7) Compute 781 × 819.

8) A gardener planted 103041 trees in such a way that the number of rows were as many as trees in a row. Find the number of rows.

9) Find the value of 511^2.

10) Packets of chalk are kept in a box in the form of a cube. If the total number of packets is 2197, find the number of rows of packets in each layer.

11) Given a + b = 10, ab = 1. Compute the value of $a^4 + b^4$.

12) The population in a village increases at a rate of 5% every year. If the population in a given year is 80000, what is the population after 3 years.

13) What is the length of the greatest rod which can fit into a room of length = 24 mt, breadth = 8 mt and height = 6 mt.

14) The difference on increasing a number by 8% and decreasing it by 3% is 407. What is the original number?

15) Rs 500 grows to Rs 583.20 in 2 years compounded annually. What is the rate of interest?

16) A number 19107 is divided into 2 parts in the ratio 2 : 7. Find the two numbers.

17) A sum of Rs 21436 is divided among 92 boys. How much money would each boy get?

18) Evaluate 1399^2.

19) Evaluate 397 * 397 + 397 * 104 + 104 * 104 + 397 * 104.

20) If the surface area of the side of a cube is 225 cm² what is the volume of the cube?

Solutions using vedic maths

1) **Which is the greatest 6-digit number which is a perfect square?**

   Solution : A 6-digit number will have a square root of exactly 3 digits.

   The largest 3-digit number is 999. Hence, the square of 999 is the required number.

   $999 \times 999 = 998001$ (Refer Chapter 1)

2) **If the length and the breadth of a square is increased and reduced by 5% respectively, what is the percentage increase or decrease in the area?**

   Solution : Let the length and breadth be 'l' and 'b' respectively. The new length will be 1.05l and the new breadth will be 0.95b. Hence, the new area will be 1.05l × 0.95 b. By nikhilam, we get

   $1.05 \times 0.95 = 100/-25 = 0.9975$

   The new area is 0.9975 lb and hence the area decreases by (1.0000 − 0.9975) which is 0.25 %. Or, the decrease can also be read off directly as the right side of the product viz 100/−25.

3) **Compute 896 × 896 − 204 × 204**

   Solution : The problem belongs to the type a² − b² where the factors are (a−b) and (a+b). Hence the solution is

   (896−204) (896+204) = 692 * 1100

                              = 761200   (Refer Chapter 3)

4) **If all the digits in numbers 1 to 24 are written down, would the number so formed be divisible by 9?**

   Solution : The number would be

123456789101112131415161718192021222324.

The sum of the first 'n' natural numbers is n(n+1)/2.

Hence, the sum would be 24 * 25 / 2 = 300 and the remainder is 3. Therefore the number is not divisible by 9.

(Refer Chapter 2)

5) **What should be the value of * in the following number so that it is divisible by 11?**
- 8287*845
- *381

Solution : For the first number, the sums of the alternate sets are 22 and (20 + *). If the value of the '*' is 2, the difference will be zero and the number will be divisible by 11. Hence, '*' = 2.

For the second number, the sums are 4 and (8 + '*'). If the value of the '*' is 7, the difference will be 11 and the number will be divisible by 11. Hence, '*' = 7.

6) **Which of the following number is divisible by 99?**
135792, 114345, 3572404, 913464

Solution : The number should be divisible both by 9 and 11.

Navasesh of each of the numbers is 0, 0, 7, 0

So, 3rd number is rejected.

Of the balance, only 2nd number is divisible by 11

(Refer chapter 3).

Hence, 114345 is divisible by 99.

7) **Compute 781 × 819.**

Solution : Use the base as 800 in Nikhilam and write as

781 – 19
819 + 19

Hence, the product will be 800/–381.

Multiply the left part by 8 to convert to the primary base

of 100, to get 6400/–381

So, borrow 4 from left to get the final answer as 6396/19.

8) **A gardener planted 103041 trees in such a way that the number of rows were as many as trees in a row. Find the number of rows.**

Solution : Compute the square root of 103041

$$
\begin{array}{c|ccccc}
 & 10 & 3 & 0 & 4 & 1 \\
 & & 1 & 1 & 0 & 0 \\
6 & & & -4 & -4 & -1 \\
 & & & 6 & 0 & 0 \\
\hline
 & & 3 & 2 & 1. & 0 & 0
\end{array}
$$

Hence, the number of trees in each row is 321.

9) **Find the value of 511^2.**

Solution : Mentally, using 500 as the base,

We get $511^2 = 522/121$.

Now, convert to the base of thousand by dividing the left part of the answer by 2.

So, the final answer is 261121.

10) **Packets of chalk are kept in a box in the form of a cube. If the total number of packets is 2197, find the number of rows of packets in each layer.**

Solution : Refer to Chapter 10, take the cube root of 2197 which is 13. Hence, each row contains 13 packets.

11) **Given $a + b = 10$, $ab = 1$. Compute the value of $a^4 + b^4$.**

Solution : Since $a + b = 10$ and $ab = 1$,

$(a + b)^2 = a^2 + b^2 + 2ab$

$a^2 + b^2 \quad = 100 - 2 = 98$

$(a^2 + b^2)^2 = a^4 + b^4 + 2\,a^2\,b^2$

$a^4 + b^4 \quad = (a^2 + b^2)^2 - 2\,a^2\,b^2$

$$= 98^2 - 2$$
$$= 9604 - 2 = 9602. \text{ (Use Nikhilam to get } 98^2$$
orally)

12) The population in a village increases at a rate of 5% every year. If the population in a given year is 80000, what is the population after 3 years.

Solution : Final population $= 80000 (1 + 5 / 100)^3$
$$= 80000 (21 / 20)^3$$
$$= 10 * 21^3$$

Now, $21^3 = $

```
    8   4   2   1
        8   4
  ─────────────────
    9   2   6   1
```

Therefore, the population is 92610.

13) What is the length of the greatest rod which can fit into a room of length = 24 mt, breadth = 8 mt and height = 6 mt.

Solution : Let the length of the greatest rod be L.
Then, $L^2 = 24^2 + 8^2 + 6^2$
$$= 576 + 64 + 36 = 676$$
Therefore, L = 26 mt.
Use techniques in Chapter 8 and 9 to get the square of 24 and the square root of 676.

14) The difference on increasing a number by 8% and decreasing it by 3% is 407. What is the original number?

Solution : Let the number be N.
Now, $(1.08 N - 0.97 N) = 407$
Hence, $0.11 N = 407$ (407 is divisible by 11)
Giving N = 3700
Therefore, the original number is 3700.

15) **Rs. 500 grows to Rs. 583.20 in 2 years compounded annually. What is the rate of interest?**

Solution : $583.20 = 500 (1 + r)^2$

$116.64 = (1 + r)^2$

By Nikhilam, we know that $116.64 = 10.8^2$

Hence, required rate of interest is 8%.

16) **A number 19107 is divided into 2 parts in the ratio 2 : 7. Find the two numbers.**

Solution : The first number would be (2 / 9) * 19107. This is a division by 9, refer to Chapter 2, we get the answer as 4246.

The second number is 19107 – 4246, use mishrank to get the second as 14861.

17) **A sum of Rs 21436 is divided among 92 boys. How much money would each boy get?**

Solution : This is a division problem, refer to Chapter 6.

```
  2
9 |  21    4    3  :  6
  |        3    3     0
  |       -4   -6    -6
  |      ─────────────────
  |       30   27     0
  |   2    3    3  :  0
```

18) **Evaluate 1399².**

Solution : Use mishrank to convert the given number to $140\bar{1}$.

Now, $140\bar{1}^2$ = $1/9/_16/\bar{2}/\bar{8}/0/1$

= 1957201

19) **Evaluate 397 * 397 + 397 * 104 + 104 * 104 + 397 * 104**

Solution : The expression is

$$(397 + 104)^2 \quad = 501^2$$
$$= 502/001 \text{ (by Nikhilam)}$$
$$= 251/001$$

20) If the surface area of the side of a cube is 225 cm², what is the volume of the cube?

Solution : The area of any one side is 225.

Therefore, each side is 15 cm. (Since the number ends with 25, square root ends with 5, and to get the first digit, $1 \times 2 = 2$)

Therefore the volume is 15^3 which is 3375 cm³.

Problems for Practice

1) A shopkeeper knows that 13% oranges with him are bad. He sells 75% of the balance and has 261 oranges left with him. How many oranges did he start with?

2) What should be the value of * in the following number so that it is divisible by 9?
- 38*1
- 451*603

3) The area of a square field is 180625 mt². If the cost of fencing it is Rs 10 per metre, how much would it cost to fence it?

4) Find the number nearest to 99547 which is divisible by 687 : 100166, 99615, 99579, 98928.

5) In a parade, soldiers are arranged in rows and columns in such a way that the number of rows is the same as the number of columns. If each row consists of 333 soldiers, how many soldiers are there in the parade?

6) One side of a square is increased by 15% and the other is decreased by 11%. What is the % change in the area?

7) Find the square root of
- 530.3809
- 0.7

8) If 1000 is invested in a bank and earns 5% interest, compounded yearly, what is the amount at the end of three years?

9) The difference in compound interest compounded yearly and simple interest at 10% on a sum for a year is Rs 25. What is the sum?

10) Compute the following : 1/35, 3 /14

11) 'A' sells a radio to 'B' at a gain of 10% and 'B' sells it to 'C' at a gain of 5%. If 'C' pays Rs 864 for it, what did it cost 'A'?

12) Find the nearest number to 77993 which is exactly divisible by 718.

13) 'A' received 15% higher than 'B' and 'C' received 15% higher than 'A'. If 'A' received 9462, which of the following did 'B' and 'C' receive?
(8218, 10032 or 8042, 10881 or 1419, 10881 or 8228, 11132)

14) Find the greatest and the least 5 digit number exactly divisible by 654.

15) Anil walks 120 metres north, 70 metres west, 175 metres south and 22 metres east. Find his distance from his starting point.

Answers
1)   1200 oranges
2)   6, 8
3)   Side = 425 mt, cost = Rs 17000
4)   99615
5)   110889 soldiers
6)   Increase of 4.035%
7)   Square Roots are 23.03, 0.8366
8)   Rs 1157.625
9)   Rs. 10000
10)   0.0285714, 0.2142857
11)   Cost is Rs 800
12)   78262
13)   'B' received 8228, 'C' received 11132
14)   10464, 99408
15)   Anil is 73 metres away.

15) Anil walks 120 metres north, 70 metres west, 175 metres south and 22 metres east. Find his distance from his starting point

Answers

1) 1200 oranges
2) 6, 8
3) Side = 425 mt, cost = Rs 17000
4) 9965
5) 110889 soldiers
6) Increase of 4.03%
7) Square Roots are 25.03, 0.8366
8) Rs 1157.625
9) Rs. 10000
10) 0.0285714, 0.2142857
11) Cost is Rs.800
12) 75202
13) 'B' received 8228, 'C' received 11132
14) 10464, 99408
15) Anil is 75 metres away